ISBN 978-0-266-95801-7
PIBN 10915579

The Commonwealth of Massachusetts.

BOARD OF GAS AND ELECTRIC LIGHT COMMISSIONERS.

REPORT

OF AN

INVESTIGATION RELATIVE TO ESTABLISHING A CALORIFIC STANDARD FOR GAS.

UNDER CHAPTER 167, GENERAL ACTS OF 1916.

BY CHARLES D. JENKINS, *Inspector of Gas.*

BOSTON:
WRIGHT & POTTER PRINTING CO., STATE PRINTERS,
32 DERNE STREET.
1917.

Chem 8409.17

PUBLICATION OF THIS DOCUMENT
APPROVED BY THE
SUPERVISOR OF ADMINISTRATION.

The Commonwealth of Massachusetts.

REPORT OF AN INVESTIGATION RELATIVE TO ESTABLISHING A CALORIFIC STANDARD FOR GAS.

Board of Gas and Electric Light Commissioners.

GENTLEMEN: — In compliance with your instructions I have the honor to submit a report on the work of my department relative to the establishment of a calorific standard for the gas of the companies under your supervision.

As a preliminary to the main body of this report it may be proper and well to say that, due to much research the conclusion has been reached throughout the scientific world that in the interest of consumer and producer the calorific standard, and not the candle-power standard, should prevail and does prevail. The reason for this change of standard is, briefly, that the great bulk of the gas now used is for heat purposes, and obviously it is of prime importance to the consumer that he be supplied with a gas which will give him the best service in view of the use he makes of it.

PLAN.

Nineteen plants were designated by the Board for experimental purposes, a list of which with a description of their equipment appears in Appendix A. These nineteen plants were free to undertake the manufacture of gas to a calorific basis instead of a candle-power basis, but otherwise complying with all the legal requirements.

The experimental period was for six months, and has been extended from time to time and is still in force.

The various plants were to make daily tests and report weekly the average calorific value and candle power, with a history of the gas, its analysis, quantities made, analyses of raw materials and such other information as would have a bearing on its quality. As a check on this work the State inspectors were to make their usual routine tests unannounced and at irregular intervals, a comparison of these single tests being made with the company reports for the week. On consultation with the inspectors modifications were to be made and experiments carried on in a spirit of co-operation, endeavoring always to improve the efficiency of the gas service.

CALORIMETERS.

For the determination of calorific value a water-flow type of calorimeter was used, the Hinman-Junker being quite generally installed at both the official testing stations and works stations. The British thermal unit adopted in the law is the quantity of heat required to raise the temperature of 1 pound, avoirdupois, of water 1 degree Fahrenheit. The science of gas calorimetry, however, was developed in the laboratory and metric measurements used, the unit being the "Calorie," which is the quantity of heat required to raise 1 kilogram of water 1 degree Centigrade; this unit is still in use in France and other countries.

At the testing stations the gas supply was taken off exactly as for candle-power determinations; in fact, some of the calorimeter gas supplies were extensions of the photometer supplies. At the works testing stations many plants used averaging tanks either of the continuous-flow type or of the gas-holder type, and they have been found to give satisfactory results.

For the official testing stations it was necessary that the inspectors should be able to control quickly the temperature of inlet water to calorimeter. After some experimenting the following installation was adopted. A No. 2 flush tank with ball cock on inlet was placed 8 to 9 feet from floor, the outlet to calorimeter flowing through a 5-inch diameter strainer of 40-mesh copper gauze. A visible waste indicator was introduced on the overflow line from inlet weir to calorimeter at the height of the eye; and it was found very useful, as the minimum amount of water could be wasted and yet a constant level insured by an easy inspection of the indicator in line with the thermometer reading lenses. The water supply to tank came through a gas tank heater. On the outlet was a brass cross (all the piping being one-half inch brass) carrying a thermometer, reading to single degrees with an open scale, and two lines, with valves, one to waste and the other to tank. On the tank line a safety valve was introduced before the valve. The temperature of water flowing to supply tank could be nicely regulated to room temperature, and was further made more uniform by a stirrer introduced into tank. When the conditions were favorable for a test the gas and water were shut off at heater, and the inlet temperature at calorimeter generally was held to within one one-hundredth of a degree Fahrenheit.

CALCULATION OF HEAT VALUES.

The form used for recording observations is shown in Appendix D and the tables for computation in Appendix E. The method of calculation has been simplified by the adoption and use of a table of co-logarithmic factors (Table C, Appendix E); these are the co-logarithms of the amount of gas actually used, corrected for temperature and total pressure, when two-tenths of a cubic foot are passed by the meter. Thus the calculation is simplified to adding the logarithms of weight of water found and the corrected difference (Table A and thermometer correction) of temperature of inlet and outlet water and the logarithmic factor from Table C; to the number corresponding to the logarithm thus found is added 1 Btu[1] for heat loss and the correction for humidity from Table E, the result being total British thermal units.

For purposes of comparing results "efficiency factors" were developed, that for coal gas being Btu's per pound of coal, and for the oil in water gas a factor was developed from the formula $B - 300\frac{(1000 - 80 A)}{1000}$, where B is the observed total heating value of the carbureted water gas, and A, the number of gallons of oil per thousand cubic feet; the values, 300 Btu's for blue water gas and 80 cubic feet gas per gallon oil, have been checked by one of the companies.[2] The oil factors are for Btu's per cubic foot from the oil used. The results for the year appear as Appendix F.

---

[1] Bureau of Standards, Circular No. 48, p. 171.
[2] Haddock, Nov. 20, 1916.

EXPERIMENTS.

In addition to installing, calibrating and checking calorimeter apparatus, and making routine tests, various experiments were carried out in the study of the effects on gas manufactured to a calorific basis. As a sequel to experiments previously made, showing the varying development of light in different burners, more particularly open burners, tests were made on the effect on candle power of gases made to a calorific basis. Experiments were also carried out on the effect of distribution on calorific value, on the effect of adding superheated steam to vertical retorts and on oil washing.

BURNERS.

Gas has been valued for many years on a candle-power basis; this was useful in the earlier days of gas lighting, as the chief use of gas was for light developed by burning the gas in some form of burners. As the art developed the so-called Argand type of burner was found to give the most light, and was used for testing at the time supervision began to take form; this type has been refined until to-day the Carpenter "Metropolitan No. 2" burner is recognized as the standard for scientific testing. In the early days some form of Argand burner was in fairly general use, as it developed considerable light from the gas; under these conditions the consumer obtained practically the same amount of light as the supervising authorities certified, but with the increasing candle power, the improvement of open burners and the more general use of gas, open burners became almost universally used, while the testing was continued with burners "that were best adapted to the gas." When incandescent gas lighting became common the discrepancy was still further increased between the light the consumer obtained and what the supervising authorities stated the gas possessed. These facts, added to the comparative crudeness of testing, made it desirable to consider some more scientific method of valuing gas. As bearing on this subject a great many tests of candle power as developed by different burners and gases were made with the testing Argand and a typical open burner; the details appear in Appendix H. Tests were made in 1910 and 1912 at routine inspections when companies were operating under a candle-power basis; in 1916 tests were made of groups of candle-power and heat-unit companies. Formerly, owing to the relatively high candle-power gas supplied, consumers could realize in open burners nearly the candle power of the gas as tested, but as the candle power, shown by official testing, dropped, the relative loss in open burners was more marked. Some tests were made of two types of open burners; with the high candle-power gases these burners had the same efficiency, but with a 14.5 candle-power mixed gas the following results were obtained: —

PRESSURE (INCHES OF WATER).	BURNER A.		BURNER B.	
	Actual Candle Power.	Candle Power per Cubic Foot.	Actual Candle Power.	Candle Power per Cubic Foot.
2.0, ...	5.1	1.08	11.0	2.01
2.5, ...	5.5	0.98	11.7	1.97
3.0, ...	5.6	0.83	13.3	1.97

The difference in light given per cubic foot between 0.83 candles and 2.01 candles is marked, especially as the Argand testing burner gave 2.9 candles per cubic foot, and the incandescent would give 18 candles per cubic foot. Nearly ten times the light per cubic foot can be obtained from this quality of gas by using it in an incandescent rather than an open burner; if an open burner must be used, certainly the best type should be chosen.

The summary follows, also a comparison in percentages of loss of candle power when using gas in an open burner: —

	1910.		1912.		1916.		Loss in Open Burners, 1912-16.	
	Argand.	Open.	Argand.	Open.	Argand.	Open.	Candle Power.	Per Cent.
Water gas, . .	19.84	19.51	20.27	19.98	16.80	9.48	10.50	52.5
Coal gas, . .	16.43	14.89	16.76	14.53	16.10	12.27	2.26	15.6
Mixed gas, . .	17.85	16.35	17.75	15.68	16.83	12.25	3.43	21.9

	1916.					
	CANDLE-POWER COMPANIES.		HEAT-UNIT COMPANIES.		LOSS IN OPEN BURNERS.	
	Argand.	Open.	Argand.	Open.	Candle Power.	Per Cent.
Water gas, . . .	16.80	9.48	14.47	8.43	1.05	11.1
Coal gas, ...	16.10	12.27	15.50	11.37	0.90	7.3
Mixed gas, . . .	16.83	12.25	15.08	10.12	2.13	17.4

DISTRIBUTION.

It is well known that the candle power developed at the works is not always delivered to the consumer in its entirety, either under low pressure for short distances or high pressure for long distances. In studying the effect of distribution of gas made to a calorific basis, high-pressure distribution was used to accentuate the effect.

Old Colony.

A test was made of the water gas supplied under high pressure by the Old Colony Gas Company during September, 1916; this was a warm period, and the gas showed comparatively little loss from compression and probably none from atmospheric conditions. The works experiments were not wholly satisfactory, as the tests were on isolated samples and not averages of gas; however, there was an apparent loss of 6.4 Btu's between the holder and after the compressor; these tests were made during the make, gas being made into the holder through the purifiers until 1 P.M. September 13. The plant was then shut down for twenty-four hours and the system supplied from this stock. In the afternoon, at the Vine Street, Weymouth, testing station (approximately 1 mile from the East Braintree works), there was a gain of 23.7 Btu's to 551.4; this represented the value of the stock of gas in the holder after being compressed and sent 1 mile. The next morning, at Whitman, 12 miles from the works, tests were made at Roberts' Tack and Nail Company after the gas had been stored since 1 o'clock of the previous afternoon and the system kept filled at 10 pounds' pressure; the results, 531.7 Btu's showed a loss of 3.6 per cent.

Btu's and the uniformity of gas, checked in the afternoon of the same day at Vine Street, Weymouth, as 552.8 Btu's. The following tables show results and analyses: —

PLACE.	Date.	B. T. U.
Works, ...	September 13, ...	534.1
After compressor, ...	September 13, ...	527.7
Testing station, ...	September 13, ...	551.4
Whitman, ...	September 14, ...	531.7
Testing station, ...	September 14, ...	552.8

Analyses (Per Cent. by Volume).

	After Compressor.	Testing Station.	Twelve Miles under 10 Pounds.
Ethylene, ...	6.77	7.04	8.88
Benzol, etc., ...	2.01	1.86	0.79
Marsh gas, ...	9.10	10.16	10.39
Hydrogen, ...	41.05	40.50	37.93
Carbon monoxid, ...	34.13	34.31	36.01
Nitrogen, ...	2.43	2.24	2.55
Oxygen, ...	0.17	0.14	0.21
Carbon dioxid, ...	4.34	3.75	3.24

Malden–Revere.

In October, 1916, and again in February, 1917, tests were made for loss of Btu's from distribution on a 6-inch pressure line between Malden and Revere, a distance of 6 miles. In the first series gas was taken from the line to compressor and holder; an average of 70 per cent. water gas for the first day was mixed with the coal gas before the station meter, but the mixture fluctuated with the coal-gas production. On the second day only 50 per cent. water gas was used. For the series of tests in February, 1917, the gas at Malden was taken after the compressor, with 30 pounds pressure, and reduced through an individual governor to 5 inches; thus the conditions, as compared with those of October 5, were less severe, and showed the effect only of distribution and weather, while the October tests showed, in addition, the effects of compression. In October the first test, on the 4th, was of gas at the usual pressure of about 8 pounds, while the second test, on the 5th, was of gas under 30 pounds, a pressure used when the occasion demanded; the flow was at the rate of 33,000 cubic feet per hour the first day, and at 65,000 cubic feet the second day. At Revere the gas for test was taken off just after the reducing governor. Under the 8 pounds' pressure there was an actual increase found in the heat value at Revere, while under 30 pounds' pressure the loss was 1.62 per cent.

B. T. U., Oct. 4, 1916, 8 Pounds Pressure.

Time.				Malden.	Revere.	Time.				Malden.	Revere.
10.00,	.	.	.	559.4	–	2.00,	.	.	.	540.0	534.9
10.30,	.	.	.	549.2	–	2.30,	.	.	.	551.6	546.7
10.50,	.	.	.	–	560.0	3.00,	.	.	.	532.6	554.5
11.00,	.	.	.	564.4	575.2	3.30,	.	.	.	507.2	535.1
11.15,	.	.	.	–	578.2	3.40,	.	.	.	–	535.6
11.30,	.	.	.	{ 535.4	} 576.3	4.00,	.	.	.	572.4	516.1
				{ 546.1		4.10,	.	.	.	–	515.9
12.00,	.	.	.	–[1]	570.3	4.15,	.	.	.	567.0	–
12.20,	.	.	.	545.9	–	4.30,	.	.	.	539.4	560.3
12.30,	.	.	.	–	563.5	4.45,	.	.	.	–	563.2
12.35,	.	.	.	538.8	–	5.00,	.	.	.	–	547.4
12.50,	.	.	.	538.8	–	5.15,	.	.	.	–	568.5
1.00,	.	.	.	533.1	552.1	Average,		.	.	544.9	552.7
1.30,	.	.	.	542.5	548.1						

[1] Cleaning pipe.

B. T. U., Oct. 5, 1916, 30 Pounds Pressure.

Time.				Malden.	Revere.	Time.				Malden.	Revere.
10.00,	.	.	.	526.5	532.1	1.00,	.	.	.	539.9	542.4
10.30,	.	.	.	556.0	562.4	1.30,	.	.	.	530.6	524.9
11.00,	.	.	.	549.0	522.8	2.00,	.	.	.	517.0	535.9
11.30,	.	.	.	586.3	558.1	2.15,	.	.	.	558.5	–
11.45,	.	.	.	543.7	–	2.30,	.	.	.	561.0	540.1
12.00,	.	.	.	566.8	534.4	Average,		.	.	550.4	541.4
12.30,	.	.	.	562.2	560.5						

Analyses (Per Cent. by Volume).

	Malden, Oct. 4, 1916.	Revere, Oct. 4, 1916.	Malden, Oct. 5, 1916.	Revere, Oct. 5, 1916.
Specific gravity, .	0.584	0.610	0.587	0.582
Ethylene, .	6.08	6.78	6.26	6.06
Benzol, .	1.69	1.60	1.08	0.87
Marsh gas, .	19.72	20.18	26.18	26.86
Hydrogen, .	38.06	36.48	35.25	35.02
Carbon monoxid, .	22.85	22.25	15.48	15.56
Nitrogen, .	6.98	7.65	10.68	10.33
Oxygen, .	0.11	0.40	0.61	0.93
Carbon dioxid, .	4.51	4.66	4.46	4.37

Similar tests were made in February, 1917, to bring in cold weather conditions; the gas at Malden was taken off just after the compressor, so that any change in the gas would come from distribution (6 miles) and cold. For some weeks the temperature had been below freezing and snow and ice were on the ground, which was frozen for $2\frac{1}{2}$ to 4 feet, the temperature at 6 P.M. February 15 being 28° F., and at 6 A.M. February 16, the day of the test, 21°. The gas (74 per cent. water gas) in the latter part of its travel passed over a railroad bridge. The loss in Btu's was 36.5, or 6.37 per cent. An interesting comparison is given later in the following tables of the results in averages reported by the two companies for three weeks, including that in which the tests were made: —

B. T. U., Feb. 16, 1917 (74 Per Cent. Water Gas).

Time.	Malden.	Revere.	Time.	Malden.	Revere.
10.00, .	571.0	-	11.30, .	559.0	-
10.07, .	570.0	-	11.35, .	-	523.2
10.15, .	591.8	-	11.45, .	552.0	527.6
10.25, .	584.6	549.1	11.50, .	561.4	-
10.30, .	572.6	559.9	12.00, .	555.1	526.6
10.45, .	539.4	533.9	12.05, .	-	524.2
10.52, .	562.0	-	12.15, .	575.1	529.9
11.00, .	619.7	514.1	12.20, .	578.1	531.6
11.05, .	606.3	543.0	12.25, .	579.3	-
11.10, .	563.0	-	12.30, .	584.9	532.1
11.15, .	558.4	568.0	12.40, .	-	535.5
11.22, .	570.0	-	Average, .	572.7	536.2
11.25, .	-	543.7			

Analyses (Per Cent. by Volume).

	Malden.	Revere.
Ethylene, ...	8.06	6.44
Benzol, ...	1.31	1.17
Marsh gas, ...	16.86	19.61
Hydrogen, ...	37.22	35.31
Carbon monoxid, ...	25.63	21.03
Nitrogen, ...	6.17	10.88
Oxygen, ...	0.35	0.81
Carbon dioxid, ...	4.38	4.75

Average Daily B. T. U. reported by Gas Companies covering Period of Distribution Tests.

DATE.	Works.	Malden Testing Station.	Suburban (Revere) Testing Station.
February 5, ...	528.0	526.0	522.5
February 6, ...	531.0	531.0	542.0
February 7, ...	544.0	538.0	540.0
February 8, ...	550.0	541.0	539.5
February 9, ...	544.0	540.0	533.0
February 10, ...	535.0	534.0	525.5
February 12, ...	536.0	535.0	528.0
February 13, ...	538.0	541.0	537.5
February 14, ...	545.0	550.0	546.0
February 15, ...	557.0	556.0	547.0
February 16, ...	563.0	561.0	548.0
February 17, ...	554.0	556.0	536.5
February 19, ...	554.0	553.0	536.5
February 20, ...	560.5	550.0	549.5
February 21, ...	560.7	566.0	553.0
February 22, ...	556.7	–	534.0
February 23, ...	550.0	548.5	546.6
February 24, ...	557.7	554.0	552.5
February 26, ...	545.5	541.0	540.5
February 27, ...	546.0	540.0	542.0
Average, ...	547.8	545.3	540.0

Taunton–Attleboro.

Taunton is a coal-gas plant, horizontals, machine-charged, making about 700,000 feet per day; of this, 75,000 cubic feet is sent to Attleboro, under an initial pressure of 20 pounds, through 1 mile of 2-inch pipe enlarging to 8 inches for 13 miles. This afforded an opportunity for testing the effect on Btu's of high-pressure, long-distance distribution of coal gas. The Taunton tests were made at the up-town testing station, while the gas for compression was taken from the same line as it left the holder. On Nov. 2, 1916, and again on March 30, 1917, when the main had been exposed to freezing temperatures during the winter and was still at about 32° F. in places, tests were made for British thermal units; in the first case little loss was shown, the average at Taunton being 586.4 and at Attleboro 585.5 Btu's. On the second test the Taunton Btu's averaged 642.4 and the Attleboro 627.6, a loss of 2.30 per cent.; this is a small loss on a high heat unit content. The samples for analysis were taken by an averaging bottle.

B. T. U., Nov. 2, 1916.

Time.					Taunton.	Attleboro.	Time.					Taunton.	Attleboro.
10.30,	.	.	.	.	590.1	–	1.15,	.	.	.	.	–	579.9
11.00,	.	.	.	.	586.8	585.6	1.30,	.	.	.	.	589.7	581.3
11.15,	.	.	.	.	–	587.1	1.45,	.	.	.	.	–	584.5
11.20,	.	.	.	.	–	584.8	2.00,	.	.	.	.	585.5	589.4
11.30,	.	.	.	.	589.4	587.1	2.15,	.	.	.	.	–	589.1
11.35,	.	.	.	.	–	584.6	2.30,	.	.	.	.	576.3	588.3
11.45,	.	.	.	.	–	587.6	2.45,	.	.	.	.	–	588.5
11.50,	.	.	.	.	–	586.9	3.00,	.	.	.	.	585.0	587.9
12.00,	.	.	.	.	590.7	585.1	3.15,	.	.	.	.	–	585.2
12.05,	.	.	.	.	–	583.5	3.30,	.	.	.	.	585.8	586.7
12.25,	.	.	.	.	–	584.3	3.45,	.	.	.	.	–	581.6
12.30,	.	.	.	.	583.8	586.1	4.00,	.	.	.	.	582.5	585.2
12.45,	.	.	.	.	–	585.6	Average,		.	.		586.4	586.0
1.00,	.	.	.	.	590.7	582.3							

Analyses (Per Cent. by Volume).

	Taunton.	Attleboro.
Ethylene, ...	2.41	2.37
Benzol, ...	0.56	0.79
Marsh gas, ...	31.33	31.77
Hydrogen, ...	50.80	51.42
Carbon monoxid, ...	8.47	7.71
Nitrogen, ...	4.10	3.68
Oxygen, ...	0.32	0.28
Carbon dioxid, ...	2.01	1.98

B. T. U., March 30, 1917.

Time.	Taunton.	Attleboro.	Time.	Taunton.	Attleboro.
10.30, ...	–	640.6	12.15, ...	–	628.7
10.35, ...	650.6	–	12.30, ...	633.9	626.6
10.40, ...	646.0	–	12.45, ...	–	627.0
10.45, ...	–	634.5	1.00, ...	647.7	625.9
11.00, ...	643.9	629.0	1.15, ...	–	621.7
11.15, ...	–	626.9	1.30, ...	646.0	624.0
11.30, ...	635.3	625.7	1.45, ...	–	620.1
11.45, ...	–	628.6	2.00, ...	642.6	–
12.00, ...	635.6	627.3	Average, . .	642.4	627.6

Analyses (Per Cent. by Volume).

	Taunton.	Attleboro.
Ethylene, ...	4.22	3.24
Benzol, ...	0.27	0.79
Marsh gas, ...	36.95	36.61
Hydrogen, ...	45.59	47.10
Carbon monoxid, ...	7.19	7.24
Nitrogen, ...	3.69	3.20
Oxygen, ...	–	–
Carbon dioxid, ...	2.09	1.82

STEAMING VERTICALS.

Vertical installations for coal-gas manufacture are of two types, intermittent and continuous; in the continuous type the coal is fed at the top, and the coke withdrawn at the bottom of retort in such manner that the carbonization proceeds continuously. This type offers a splendid opportunity, in theory, of "improving the art of gas manufacture" by increasing the heat units per pound of coal by the introduction of steam. Experiments were carried out at the Springfield Gas Light Company's plant in March, 1917, with satisfactory results.

Unit B, consisting of 24 retorts, Glover-West, in three benches of eight retorts each, was so arranged that the gas could be purified and kept separate from the other gases until a sample could be taken in the averaging holder. After experimenting on amount of steam that could be used, the following method was decided on: 20 pounds' pressure steam was fed through a fixed aperture, reducing the pressure to 9–9.5 pounds; this steam was passed through two lengths of pipe placed in a waste heat flue and acquired about 200° F. superheat; then passing through an aperture less than two-tenths inch diameter the steam entered the retort at its lower end at the rate of 350 pounds per twenty-four hours per retort. The superheated steam passing through the red-hot coke in lower part of retort would be partially decomposed, forming blue water gas of approximately equal parts carbon monoxid and hydrogen; but analysis of the gas shows further action of the steam and water gas in preventing the cracking of the hydrocarbons. The hydrogen in the steamed sample should have increased by the same amount as the carbon monoxid if the action were only that of making blue water gas, but as the hydrogen remained the same, in proportion, as before steaming, while the carbon monoxid was increased (over 3 per cent.), and the illuminants increased in total amount and also in proportion, it is fair to deduce that the introduction of steam improved the quality of gas produced as well as its quantity. In addition to this another advantage appeared in the changed character of carbon in retorts; generally this is in a hard, com-

pact form detached with difficulty from the retort, but when
steamed the carbon is more amorphous, is more easily detached by barring and burns off quicker; in fact, the tests
showed 43.7 per cent. less time lost from retorts out for scurf-
ing. There is also an apparent increase in the ammonia production of over 12 per cent.

The actual tests ran two weeks, after considerable preliminary work had been done. Coal of at least three varieties, suf-
ficient for the two weeks' test, was well mixed in the storage
shed to insure a uniform supply. A 15-foot sampling holder
was so arranged as to collect an average sample of the gas for
the day period from 8 A.M. to 5 P.M., and for the night period
from 5 P.M. to 8 A.M. The company's chemist made tests on
each day and night sample, and the State inspectors made one
day and night sample test each week for a check. The first
week was run without steam, the second week with steam.
The results follow: —

SPRINGFIELD STEAMING TESTS.

Company's Averages for Each Week.

	Not steamed.	Steamed.	Increase (Per Cent.).
Average daily make (cubic feet),[1] . . .	650,000	730,000	12.30
Average yield per pound (cubic feet), . .	5.49	6.72	22.40
Average Btu, ...	540	519	3.89[2]
Average Btu per pound, ...	2,965	3,488	17.60
Average candle power,[3]	9.5	7.0	26.30[2]
Average retorts out per day, . . .	1.81	1.02	43.70[2]

[1] One week without and one week with steaming.   [2] Decrease.   [3] Argand burner.

State Inspectors' Results and Analyses.

	March 16, 1917, not steamed.	March 23, 1917, steamed.	
		Day.	Night.
Average Btu total, ...	523.40[1]	516.00	511.20[2]
Average Btu net, ...	475.70	468.80	463.20
Candle power, ...	8.27[3]	6.10	6.40[4]
Yield (feet per pound), ...	5.49	6.72	6.72[5]
Btu per pound of coal, ...	2,873.00[6]	3,466.00[7]	3,435.00[8]

Analyses (Per Cent. by Volume).

Ethylene, ...	1.07	1.79	1.51
Benzol, ...	0.97	0.84	1.06
Marsh gas, ...	27.33	24.93	24.42
Hydrogen, ...	57.69	56.11	56.25
Carbon monoxid, ...	7.92	11.03	11.34
Nitrogen, ...	3.80	3.55	3.79
Oxygen, ...	–	–	–
Carbon dioxid, ...	1.23	1.75	1.63

[1] Company's average for the week, 540.
[2] Company's average for the week, 519.
[3] Company's average for the week, 9.5.
[4] Company's average for the week, 7.0.
[5] 19.57 per cent. increase.
[6] Company's average for the week, 2,965.
[7] 20.65 per cent. increase.
[8] Company's average for the week, 3,488.

The results show an increased make of gas per pound of coal, an increased "efficiency" factor, a better quality of gas, an increased yield of ammonia, less wear and tear on the retorts and no disadvantages as yet.

OIL WASHING.

Having in mind the object to improve the efficiency of service, and knowing the difficulties encountered in delivering gas uniform in quality at the consumer's premises, the question of distribution troubles and their remedies was studied. One difficulty is the presence in the commercial gas of naphthalene and easily condensible hydrocarbon vapors; these are affected by size of mains, speed of flow of gas, fluctuations in pressure and weather conditions, at times dropping out of the gas and at other times being picked up by the gas. On a candle-power basis it is desirable to have the gas carry as much of these elements as it will, but on a heat-unit basis they are not so desirable. Naphthalene can be reduced by washing the gas with various oils; this washing process may reduce the amount of lighter vapors and olefine gases that are not so undesirable. Washing gas with oil is the usual method of recovering the benzol, toluol, etc., carried by the gas, and experiments were instituted and much study given to the effect on its heating value by so doing. Various oils may be used, the usual ones being green and creosote oils, distillates from coal tar and straw and mineral oils, "non-viscous neutrals," distillates of petroleum. Washing with these oils takes from the gas some of its constituents which are recovered from the wash oils by distillation as "light oils." These light oils are composed of benzol, toluol, solvent naphthas and other heavier vapors, as well as naphthalene. At present toluene (the pure toluol) is of importance as a base for the manufacture of high explosives, and, owing to the difficulty of getting apparatus for its recovery from gas, study was given to this phase of the question. The problem became one of improvising apparatus to wash gas and to recover the light oils. Gas oil as used for making water gas was used, in a Standard washer scrubber, as a washing medium; a steam-heated still was made from a boiler feed water heater, and a condenser from a section of hydraulic main put on end with piping, later supplanted by a coil of lead pipe, for condensing the vapors. The outlet from still was a 6-inch pipe, 8 feet high, filled with stones, acting as a dephlegmator. Pipe lines, pumps and storage tanks completed the equipment. As considerable apprehension was felt by the plant manager as to naphthalene being sent out, attention was given to chemical control of this element; a method for rapid analysis was devised and tested, and it was found that, with the coal gas experimented with, the naphthalene content was reduced from between 26 to 28 grains per 100 cubic feet to between 5 to 6 grains per 100 cubic feet by washing with gas oil. An additional advantage was found in the reduction of total sulphur in the gas; before the washing the amount was 28.4 grains per 100 cubic feet, and after washing, 19.5 per 100 cubic feet.

The oil used assayed 5½ per cent. off at 200° C.; 28½ per cent. off at 250° C.; 62 per cent. off at 300° C.; 88½ per cent. off at 350° C.; 11½ per cent. as residue.

A test was carried out, washing 1,000,000 cubic feet of gas with 1,556 gallons of oil in twelve hours, obtaining 74 gallons light oils, equivalent to 4.76 per cent. of the wash oil used; these light oils distilled off from the benzolized oil gave 34 per cent. to 100° C.; 22 per cent. to 120° C.; 7 per cent. to 135° C.

On fractionating with a column still head there was found 29.5 per cent. benzene, 12.8 per cent. toluene and 5.8 per cent. solvent naphthas.

The effect on the gas is shown in the following table. The retorts were charged every two hours, 24 at each charging, requiring forty minutes. The oil was pumped into washer scrubber beginning at 9 o'clock and occupying nearly an hour. The tests were made on gas from an averaging holder which showed a lag of thirty minutes before showing the effects of any change in the gas. Therefore the first test would be of gas before the effects of charging and washing were felt.

Time.		B. T. U.	Candle Power.
9.15 A.M.,	.	597.8	13.75
9.45 A.M.,	.	599.2	13.66
10.15 A.M.,	.	604.7	14.23
10.45 A.M.,	.	595.8	14.18
11.15 A.M.,	.	587.7	14.01
11.45 A.M.,	.	581.2	13.03
12.15 P.M.,	.	584.5	12.33
1.00 P.M.,	.	587.1	12.04
1.30 P.M.,	.	584.9	12.13
2.00 P.M.,	.	578.4	12.55
2.30 P.M.,	.	598.8	13.05
3.15 P.M.,	.	602.3	13.47
Average,	.	591.8	13.20

CONCLUSIONS.

In the report of the gas inspector for the year 1895 [1] appear the first calorific determinations of Massachusetts gases. Considerable work has been done since by various committees, gas associations and the Bureau of Standards in perfecting apparatus and refining methods, until to-day calorific value is generally recognized as the most scientific method of valuing gas.

In 1915 a series of tests was made [2] covering the typical gases supplied by the Massachusetts companies, — coal gas, water gas and mixed coal and water gases. If the average candle power found in these tests be calculated to 16, the minimum required by the statutes, and the average calorific value be reduced proportionately, there would be developed a "translation value" of 560.9 Btu's.; that is, if all the gases tested had been exactly 16 candle power the equivalent in calorific value would have been 560.9 Btu's. It does not follow, however, that if gas should be made to a calorific value of 560.9 Btu's, the candle power would be 16. This investigation was undertaken, in part, precisely for such information, and the averages for the several periods show the results obtained.

Averages, Weekly Reports.

PERIOD OF TIME.	B. T. U.	Candle Power.
July 1, 1916, to Jan. 1, 1917, ...	559.7	15.29
Jan. 1, 1917, to Apr. 1, 1917, ...	558.7	13.64
Apr. 1, 1917, to July 1, 1917, ...	557.9	14.50
July 1, 1917, to Oct. 1, 1917, ...	556.7	14.63
July 1, 1916, to Oct. 1, 1917, ...	558.5	14.83

Even under the stress of abnormal times the calorific value has remained exceedingly constant, while the candle power has followed the usual fluctuation experienced in changing weather. Uniformity of calorific value is to be desired, for the larger proportion of gas is used as a source of heat, and a uniform quality of gas is the first element of good service to the con-

---

[1] Public Document No. 55, January, 1896.
[2] Thirty-first annual report, Board of Gas and Electric Light Commissioners.

sumer. If, in order to retain a uniform candle power, the calorific value varies, the actual service rendered the majority of
consumers will be impaired.

Good service consists of uniform quality of gas supplied
under uniform pressure through efficient appliances kept prop-
erly adjusted. The gas should be of such quality that the consumer gets the greatest amount of useful, available heat at the
least cost. This means that the diluents, or inert gases, should
be kept as low as commercially practicable, and that the gas
should be so made as to insure the delivery to all consumers,
under all conditions of distance and weather, of the largest pos-
sible amount of the heat units produced in the manufacture of
the gas.

Experiments already described show the loss of calorific value
from high pressure and long-distance distribution to be from
2.3 to 6.5 per cent. Adding steam to vertical retorts increases
the "efficiency factor" 20.1 per cent., as well as being advan-
tageous in other respects. Tests show that open burners do not
develop the full lighting value of gas as determined by the official burner; the loss is from 7.3 per cent. for coal gas to 52.5
per cent. for water gas. Open burners vary widely in efficiency
among themselves, the tests ranging from 28.6 to 69.3 per cent.
of the true lighting value of the gas. The oil-washing experiments show the possibility of separating benzol, toluol, etc.,
from gas by washing with oil in improvised apparatus. While
oil-washing "hardens" the gas and tends to increase uniformity
of quality by removing condensible hydrocarbon vapors, naphthalene and sulphur, it reduces also the calorific value somewhat.
As toluol is a base for the manufacture of high explosives, and
is needed by the government for such use, and as benzol may
also be so needed, it is worth noting that the removal of
these elements would reduce the calorific value by 6 to 7 per
cent. only. But as the urgent need of the Federal government
for toluol necessitates the recovery of this element from gas, the
minimum calorific value required of the gas companies should
be set at such a figure as to make this recovery feasible. Con-
sidering the results of this investigation, the uncertainty of a
supply of suitable coal and oil for gas manufacture, and the
need of the Federal government for toluol, and possibly, later,

benzol, it would seem desirable for the Board to establish a minimum calorific standard of not over 525 total British thermal units, and to recognize the possibility that this figure may need revision in the early future.

The appendices consist of the following: —

A. List of companies participating and their plant equipment.
B. Diagram of calorimeter tank installation.
C. Form used for weekly report by the gas companies.
D. Form for recording observations and calculating results of calorific determinations.
E. Tables for calculation.
F. Efficiency factors for coal and oil.
G.   I. Detailed reports Btu and candle power by the companies to Oct. 1, 1917.
     II. State inspections and comparisons.
     III. Curves Btu for year ending June 30, 1917.
H. Burner tests, Tables A to I.

Respectfully submitted,

CHAS. D. JENKINS,
Inspector.

Nov. 12, 1917.

APPENDIX A.

LIST OF COMPANIES PARTICIPATING, AND THEIR PLANT EQUIPMENT.

COMPANY.	Equipment.
Attleboro (No. 15),	Coal gas (buying 25-30% from Taunton). Seven benches of 6's; four 10' x 15" x 30" and three 9' x 14" x 26". Capacity 0.3 million. Daily tests at testing station; no calorimeter at works.
Boston (No. 1),	Mixture coke oven and water gases in about equal parts. Tests made at Everett on water gas three times daily. Three tests daily at testing station at the Central Station, Roxbury.
Brockton (No. 2),	Coal gas with 40-60% water gas. Four stacks of 4's Woodall-Duckham verticals with a capacity of 0.9 million per 24 hours. One 9' and one 7' 6" U. G. I. water gas sets, capacity 2.0 millions. Three tests daily of coal, water and mixed gases; one test daily at testing station.
Cambridge (No. 3),	Coal gas with 25-35% water gas. Twenty benches of 8's horisontals, 20' x 16" x 20", capacity 3.5 millions. Two 8' 6" sets U. G. I. and one 8' 6" set Western Gas Construction Company, water gas, capacity 3.5 millions. Tests made at works four times daily from sampling tanks of coal and water gases; two tests street gas. One test daily at testing station.
Charlestown (No. 4),	Coal gas. Nine benches of 8's horisontals, 20' x 15" x 26". One test daily testing station; no calori·meter at works.
East Boston (No. 19),	Coal gas with 40-50% water gas. Tests made once daily at office; no calorimeter at works.
Fall River (No. 5),	Coal gas with about 50% water gas. Seven stacks of 9's U. G. I. verticals, capacity 1.05 millions. Four sets 8' 6" water gas, capacity 4.0 millions. Tests on coal gas from averaging (24 hours) holder, 1 Btu and candle power every 2 hours; water gas, snap test Btu and candle power every hour. Commercial gas, one test daily at works and testing station.
Fitchburg (No. 6),	Coal gas with about 50% water gas. Two banks of 8's Glover-West verticals, capacity 0.4 millions. One 6' U. G. I. and one 6' 6" Lowe gas machine for water gas, capacity 0.9 million. Tests on coal gas twice daily from sampling holder; three times on water gas. Commercial gas twice daily leaving holder, and once at testing station.
Haverhill (No. 7),	Straight water gas. Two 8' 6" U. G. I. sets, capacity 2.0 millions. Tests made three times daily on main to holders, and once at testing station.
Holyoke (No. 8),	Coal gas with 20-25% water gas. Six stacks of 4's Woodall-Duckham, capacity 0.75 million. One 7' 4" set Humphreys & Glasgow and 7' 6" set U. G. I. water gas, capacity 1.25 millions. Daily tests on 24 hours averaging holder sample, and two, morning and afternoon, on coal gas; two tests on water gas. One test at testing station.
Lowell (No. 9),	Coal gas with about 70% water gas. Ten benches of 12's horisontals 14' long, capacity 1.6 millions. One 8' 6" and one 9' Lowe water gas sets, capacity 3.2 millions. Tests once daily, coal, water and commercial gases, the first two from averaging tanks. Once daily testing station.
Lynn (No. 16),	Coal gas with 40-50% water gas. Ten benches of 9's inclined silica 18' x 15" x 26", capacity 1.8 millions. Two 8' 6" U. G. I water gas sets, capacity 4.2 millions. Tests on each gas once daily, and once on commercial gas at testing station.

24

LIST OF COMPANIES PARTICIPATING, AND THEIR PLANT EQUIPMENT
— Concluded.

COMPANY.	Equipment.
Malden (No. 10), . . .	Coal gas with 50–60% water gas. Five stacks of 9's inclined, capacity 1.0 million. One 11′, one 9′ and one 7′ 6″ water gas sets, capacity 5.0 millions. Tests on coal gas, from sampling tank, twice daily; water gas three times daily. Two tests daily on commercial gas leaving holder, and one at testing station.
New Bedford (No. 11), .	Coal gas with 38% water gas. Six benches of 6's inclined, 20′ x 26″ x 16″, capacity 0.57 million. Water gas, two 9′ U. G. I. and one 7′ 6″ Western Gas Construction Company, capacity 3.15 millions. Averaging tanks used for both coal and water gas; one test of each gas daily, with a Parr calorimeter. One test of commercial gas at testing station.
Old Colony (No. 12), . .	Straight water gas distributed entirely under high pressure (8–30 pounds). One 7′ and one 6′ U. G. I. sets. Daily tests at testing station.
Springfield (No. 13), .	Coal gas with 50–60% water gas. Five sets of 9's inclined and six sets of 8's Glover-West verticals, capacity 2.5 millions. Two 9′ and one 10′ U. G. I. water gas sets, capacity 4.0 millions. Tests on mixed coal gases from averaging holder made twice daily; water gas, three tests; commercial gas, twice daily on leaving holders, and once at testing station.
Suburban (No. 17), . .	High-pressure gas from Malden, 6 miles under 5–38 pounds' pressure. Tests daily on commercial gas.
Taunton (No. 18), . .	Coal gas with 10–15% water gas when needed. Six benches 20′ x 16″ x 26″ horisontals, capacity 0.9 million. One 7′ water gas set, 0.5 million capacity. Daily tests at testing station.
Worcester (No. 14), . .	Coal gas with 40–50% water gas. Two sets of ten banks of 8's and one set eight banks of 16's horisontals, capacity 3.8 millions. One 8′ 6″ and two 8′ water gas sets, capacity 3.0 millions. Tests once daily on coal gas from 40-gallon continuous sampling tank. Water gas once a day from relief holder. Commercial gas once a day at works and at testing station.

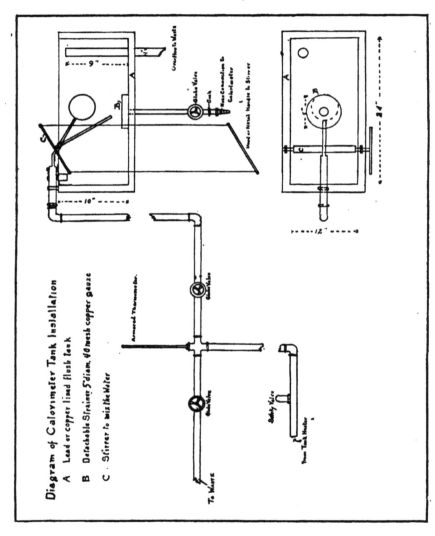

APPENDIX C.

Form G.L. B. 4,000. 4-17.

The Commonwealth of Massachusetts
CALORIFIC INVESTIGATION, WEEKLY REPORT
BOARD OF GAS AND ELECTRIC LIGHT COMMISSIONERS

Company _____ Six day period ending _____   A.M.
                                                                  P.M.   1917.

COAL GAS		WATER GAS	
Gas made° per pound of coal		Gas made° [1]	
Kinds of coal used		Generator coal lbs. per M gas	
Percentage of each		Generator coke lbs. per M gas	
		Total fuel lbs. per M gas	
Analysis each cargo above coals			
Date		Oil, gals. per M cu. ft. gas°	
Moisture	Ash		
Volatile	Sulphur	No. hours run, each machine	
Fixed carbon	B. T. U.	Tar, gals. per M cu. ft. gas	
Kind of enricher used			
Amount per hundred lbs. of coal		Kind of oil	
		Analysis of oil	
Lbs. coke per 100 lbs. coal		Date	
Gals. tar per 100 lbs. coal		Spec. gravity	
Lbs. NH, per 100 lbs. coal		Distillation up to 100° C	200 — 250
		100 — 150	250 — 300
		150 — 200	300 — 350
B. T. U. gas, average		B. T. U. gas, average	
B. T. U. gas, per lb. coal		B. T. U. gas, per gal. oil	
Candle-power	burner	Candle-power	burner
Specific gravity		Specific gravity	
Analysis gas		Analysis, or, if not complete, CO_2	
Date		Date	
Illum.		Illum.	
CH_4	N_2	CH_4	N_2
H_2	O_2	H_2	O_2
CO	CO_2	CO	CO_2

COMMERCIAL GAS

	coal		water		% water
Amount gas made° total [1]					
B. T. U. at works,[2] average	max.	date	min.	date	
B. T. U. testing station, average	max.	date	min.	date	
B. T. U. State inspection [3]		date			

Candle-power, average _____ at _____ station, burner used _____

Gas analysis [4]		State inspection [5]	Total sulphur per 100 cu. ft.
Date			average
Illum.	N_2		max. and date
CH_4	O_2		min. and date
H_2	CO_2		
CO	Spec. gravity		

REMARKS :

Signed

° All gas returns to be corrected to 30" bar. and 60° F., purified and metered   [1] Adhere to schedule; if changed, note in remarks space
[1] Thousands and tenths; Sunday make excluded   [2] Do not fill in   [4] Average, if more than one

APPENDIX D.

Form G. L. 19. 10,000. 8-'17.

The Commonwealth of Massachusetts
BOARD OF GAS AND ELECTRIC LIGHT COMMISSIONERS
DAILY CALORIFIC TESTS

						a.m. p.m.
Place		Date		Time		

Kind of Gas | No. of Calorimeter | Maker

Temp. of Room   F.   Water Valve
Humidity   %   Exhaust Valve
     Wet Bulb   Rate per Hour
     Dry Bulb

Temp. of Meter	WATER TEMPERATURE					
Meter Outlet Pressure	In	Out	In	Out	In	Out
Equivalent in Mercury						
Barometer						
Total Pressure						
Temp. of Exhaust						
Condensation; Water c.c.						
Meter start						
" end						
B. T. U. from Cond.						
GAS 1						
Meter start						
" end						
2						
Meter start						
" end						
3						
Meter start						
" end						
Av. Temp. observed						
Ther. and Stem. Corr.						
Corrected Temp.						
Rise in Temp.						
Weight of Water						
Observed B. T. U.						
Humidity Correction						
Heat Loss						
Total B. T. U.						
Average B. T. U.						
Net B. T. U.						

Remarks:

Appendix E.

Tables for Calorimetric Corrections.

Table A. — *Stem Correction for Outlet-Water Thermometer.*

Thermometer immersed to	Temperature Rise of Water.	Room Temperature.					
		50°	60°	70°	80°	90°	100°
40°	10, ...	+0.01	+0.02	+0.03	+0.03	+0.04	+0.05
	15, ...	.03	.04	.05	.06	.08	.09
	20, ...	.04	.05	.07	.09	.11	.12
50°	10, ...	+0.01	+0.01	+0.02	+0.03	+0.04	+0.05
	15, ...	.02	.03	.04	.05	.07	.08
	20, ...	.02	.04	.06	.07	.09	.11
60°	10, ...	+0.00	+0.01	+0.02	+0.02	+0.03	+0.04
	15, ...	.00	.01	.03	.04	.05	.06
	20, ...	.00	.02	.04	.05	.07	.09

Table B. — *Conversion Inches Water into Inches Mercury.*

Inches Water.	Inches Mercury.	Inches Water.	Inches Mercury.	Inches Water.	Inches Mercury.	Inches Water.	Inches Mercury.
1.0	0.07	2.4	0.18	3.8	0.28	5.2	0.39
1.2	0.09	2.6	0.19	4.0	0.30	5.4	0.40
1.4	0.10	2.8	0.21	4.2	0.31	5.6	0.41
1.6	0.12	3.0	0.22	4.4	0.33	5.8	0.43
1.8	0.13	3.2	0.24	4.6	0.34	6.0	0.44
2.0	0.15	3.4	0.25	4.8	0.36	0.1	0.0074
2.2	0.16	3.6	0.27	5.0	0.37		

TABLE D. — *Percentage of Humidity (Rapid Ventilation).*

Dry bulb reading. Degrees Fahrenheit.

88	90
	94
94	86
85	78
78	71
70	64
64	58
57	53
52	46
46	42
40	36
35	31
29	26
25	23
21	17
16	13
12	9

Total pressure, inches.

Humidity.

T₁

	Temperᴀ	
	of V	
	40° {	10,
		15,
		20,
Thermometer Immersed	S 50° {	10,
		15,
		20,
	60° {	10,
		15,
		20,

T₁

Inches Water.
1.0
1.2
1.4
1.6
1.8
2.0
2.2

TABLE D. — *Percentage of Humidity (Rapid Ventilation).*

Dry bulb reading. Degrees Fahrenheit.

	60	62	64	66	68	70	72	74	76	78	80	82	84	86	88	90
88																94
86															94	86
84														93	85	78
82													93	85	78	71
80												93	83	77	70	64
78											93	84	76	70	64	58
76										92	83	76	69	63	57	53
74									92	83	76	68	63	56	52	46
72								92	83	75	68	63	55	50	46	42
70							92	82	74	67	62	54	49	44	40	36
68						90	82	73	66	60	54	47	44	38	35	31
66					90	82	73	65	58	53	47	43	37	33	29	26
64				90	80	72	64	57	52	46	42	36	33	27	25	23
62			90	80	72	63	57	50	44	39	35	30	27	24	21	17
60		89	79	70	62	55	48	43	37	33	28	25	23	18	16	13
58	89	79	70	62	54	47	42	36	32	27	24	20	16	15	12	9
56	79	69	61	53	45	40	35	30	25	23	18	15	14	10		
54	69	60	52	45	38	33	27	24	20	16	13	.10				
52	58	49	43	37	31	25	22	17	14	11						
50	49	41	35	30	23	20	15	12								

Humidity.

TABLE E. — *Corrections for Humidity in B. T. U.*

Percentage of humidity.

	10%	20·	30	40	50	60	70	80	90
60°	+4	+4	+3	+2	+2	+1	+1	0	0
65	+5	+4	+4	+3	+2	+2	+1	0	-1
70	+6	+5	+4	+3	+3	+2	+1	0	-1
75	+7	+6	+5	+4	+3	+2	+1	0	-1
80	+8	+7	+6	+5	+3	+2	+1	0	-1
85	+10	+9	+7	+6	+4	+3	+2	0	-1
90	+12	+10	+9	+7	+5	+4	+2	0	-2

Room temperature.

APPENDIX F.

EFFICIENCY FACTORS FOR COAL AND OIL, JULY, 1916, TO OCTOBER, 1917.

COMPANY.	B. T. U. Feet.	Oil Factor.
Boston, ...	-	109
Brockton, ...	3,289	112
Cambridge, ...	3,176	105
Charlestown, ...	3,209	-
Fall River, ...	3,146	109
Fitchburg, ...	3,664	106
Haverhill, ...	-	110
Holyoke, ...	2,826	111
Lowell, ...	3,201	100
Lynn, ...	2,962	107
Malden, ...	2,666	106
New Bedford, ...	3,002	105
Old Colony, ...	-	109
Springfield, ...	2,913	101
Taunton, ...	2,980	-
Worcester, ...	2,882	104

APPENDIX G.

I. Tabulation of weekly reports and averages made by gas companies, showing averages by weeks of quality of commercial gas at testing stations and of the constituent gases at works stations; for the period from July 1, 1916, to October 1, 1917.

II. Inspection results by State inspectors and company averages reported for week during which the official test was made for the same period.

III. Curves Btu's for year ending June 30, 1917.

TABULATION OF WEEKLY REPORTS AND AVERAGES.

Attleboro.

The only results reported are of Btu for commercial gas, as follows: —

Commercial B. T. U.	Commercial B. T. U.	Commercial B. T. U.	Commercial B. T. U.	Commercial B. T. U.	Commercial B. T. U.	Commercial B. T. U.
624	611	611	610	609	634	608
622	610	606	609	611	622	598
605	608	604	623	625	633	
611	611	607	618	617	603	614
607	613	613	608	630	618	
598	618	613	597	622	614	
602	607	626	602	617	636	
594	603	632	621	626	620	

Boston.

COAL.				WATER.				COMMERCIAL.		
B.T.U.	Candle Power.	Yield.	B.T.U. Feet.	Gals.	B.T.U.	Factor.	Candle Power.	Per Cent. H_2O.	B.T.U.	Candle Power.
640	17.1	–	–	2.92	538	106	16.2	50.8	595	16.7
638	17.3	–	–	3.02	535	102	15.9	50.0	586	16.3
645	17.2	–	–	3.01	538	102	15.9	51.9	585	16.3
636	17.2	–	–	2.77	521	104	14.8	52.3	582	16.3

Boston — Continued.

COAL.				WATER.				COMMERCIAL.		
B.T.U.	Candle Power.	Yield.	B.T.U. Feet.	Gals.	B.T.U.	Factor.	Candle Power.	Per Cent. H₂O.	B.T.U.	Candle Power.
623	17.3	–	–	2.75	517	101	15.5	56.3	564	16.3
625	17.2	–	–	2.66	518	106	15.8	57.4	563	16.2
617	17.4	–	–	2.66	512	104	15.5	64.6	568	16.2
616	17.1	–	–	2.69	519	105	15.4	59.4	564	16.1
620	17.1	–	–	2.65	513	108	15.5	59.6	576	16.1
655	17.1	–	–	2.68	516	105	15.4	58.2	587	16.1
625	17.1	–	–	2.57	526	109	15.4	59.0	576	16.0
620	17.0	–	–	2.82	540	109	15.4	59.8	581	15.9
630	17.0	–	–	2.89	534	105	15.9	59.9	583	16.1
629	17.1	–	–	2.92	545	108	16.1	60.2	584	16.4
616	16.9	–	–	2.92	546	108	16.7	60.3	589	16.6
641	16.9	–	–	3.07	548	105	16.5	60.5	589	16.4
657	17.3	–	–	2.94	552	109	17.0	59.6	595	17.0
648	17.1	–	–	2.82	538	108	17.0	59.1	587	17.0
630	17.2	–	–	2.80	536	108	16.7	62.6	573	16.9
613	16.9	–	–	3.25	565	106	17.0	61.5	585	16.9
601	17.1	–	–	3.09	557	107	17.4	57.0	593	17.0
612	17.2	–	–	2.67	533	113	17.1	58.0	579	17.0
613	17.1	–	–	2.80	544	111	17.0	59.8	572	17.0
627	17.2	–	–	2.74	537	111	16.9	59.4	581	17.0
624	17.1	–	–	2.77	537	111	16.9	59.4	579	17.0
628	17.0	–	–	2.84	537	107	16.9	60.0	573	16.9
611	16.9	–	–	3.13	559	107	17.2	62.6	578	17.0
624	17.1	–	–	2.96	543	106	17.1	61.2	574	16.9
634	17.1	–	–	2.83	543	110	17.1	58.9	577	17.1
632	17.2	–	–	2.81	537	108	16.5	59.6	572	16.8
627	17.1	–	–	2.86	540	108	16.5	58.4	573	16.7
634	17.3	–	–	2.67	526	109	16.2	55.5	578	16.8
638	16.9	–	–	2.77	543	108	16.1	60.3	568	16.4
630	17.3	–	–	2.82	537	108	16.7	59.2	580	17.0
628	16.6	–	–	2.48	527	115	15.4	58.3	579	16.0
638	17.1	–	–	2.57	530	114	15.4	56.5	586	16.2
655	17.2	–	–	2.65	535	113	15.7	56.5	580	16.4
654	17.0	–	–	2.69	535	112	16.0	56.4	589	16.4
637	16.8	–	–	2.57	543	119	16.4	58.0	587	16.8

Boston — Concluded.

COAL.				WATER.				COMMERCIAL.		
B.T.U.	Candle Power.	Yield.	B.T.U. Feet.	Gals.	B.T.U.	Factor.	Candle Power.	Per Cent. H₂O.	B.T.U.	Candle Power.
630	17.1	–	–	2.58	534	115	16.2	56.9	582	16.7
643	17.1	–	–	2.54	542	119	.16.1	60.2	579	16.7
652	17.1	–	–	2.74	539	111	15.8	58.3	584	16.5
629	16.6	–	–	2.55	531	116	15.5	58.8	579	16.1
636	16.6	–	–	2.58	538	116	15.5	59.5	591	15.9
652	16.1	–	–	2.67	544	115	15.6	57.8	589	15.7
643	17.0	–	–	2.59	543	118	16.2	57.4	595	16.5
657	17.0	–	–	2.66	550	118	16.1	58.9	593	16.6
655	17.2	–	–	2.62	537	115	15.8	54.2	595	16.6
641	17.1	–	–	2.62	529	112	15.6	54.3	583	16.3
654	17.0	–	–	2.62	529	112	15.4	52.3	590	15.7
635	15.9	–	–	2.61	535	114	14.9	53.4	584	15.2
623	15.3	–	–	2.62	539	115	14.5	44.9	589	14.8
605	15.2	–	–	2.60	532	113	14.9	53.5	569	14.7
609	15.1	–	–	2.65	537	114	15.2	55.6	569	14.8
620	15.2	–	–	2.83	555	114	16.2	53.8	586	15.5
606	15.0	–	–	2.69	548	116	15.8	54.9	576	15.2
608	14.1	–	–	2.69	529	110	15.3	56.1	567	14.7
601	13.8	–	–	2.66	529	110	14.7	59.5	558	13.9
594	13.6	–	–	2.74	540	111	15.1	56.5	564	14.1
582	13.4	–	–	2.72	548	115	15.2	57.5	561	13.8
629	16.6	–	–	2.75	537	109	16.0	57.5	580	16.2

Brockton.

COAL.				WATER.				COMMERCIAL.		
B.T.U.	Candle Power.	Yield.	B.T.U. Feet.	Gals.	B.T.U.	Factor.	Candle Power.	Per Cent. H₂O.	B.T.U.	Candle Power.
–	15.0	5.60	–	2.94	–	–	17.6	56.0	567	15.1
–	15.4	5.41	–	3.00	–	–	18.0	54.0	567	–
574	15.6	5.35	3,071	3.14	566	103	18.6	55.0	574	14.8
571	15.2	5.87	3,352	3.00	584	119	19.0	59.0	565	16.0
581	13.7	5.98	3,474	2.83	558	115	17.2	54.0	558	15.9
568	14.5	5.83	3,311	2.89	555	111	17.0	50.2	554	15.9
570	14.9	5.99	3,414	2.94	568	115	17.4	50.0	553	16.0
573	14.2	6.06	3,484	2.66	570	102	16.7	45.0	561	15.8
581	15.5	5.54	3,219	2.76	553	116	16.1	47.5	555	16.0
582	13.7	5.76	3,352	2.84	560	116	16.6	44.7	555	15.6

Brockton — Continued.

COAL.				WATER.				COMMERCIAL.		
B.T.U.	Candle Power.	Yield.	B.T.U. Feet.	Gals.	B.T.U.	Factor.	Candle Power.	Per Cent. H_2O.	B.T.U.	Candle Power.
598	14.5	5.89	3,522	2.65	556	121	17.4	48.1	553	15.0
575	13.2	5.94	3,416	2.66	557	121	15.7	48.3	549	14.3
589	14.1	5.83	3,434	2.90	564	115	17.1	48.2	554	14.2
592	13.9	6.06	3,588	2.93	560	112	16.7	51.6	559	13.5
595	14.5	5.82	3,463	2.86	552	112	17.0	51.4	563	14.1
587	13.3	5.55	3,434	2.93	549	109	16.2	49.0	559	13.3
591	14.6	5.74	3,392	2.90	597	127	16.6	55.3	553	13.5
604	14.5	5.65	3,413	2.99	551	108	16.6	52.0	555	13.7
601	14.6	5.59	3,360	3.01	571	116	18.0	53.4	555	13.7
589	14.1	5.90	3,475	2.66	567	124	17.1	57.4	551	13.5
596	14.5	5.57	3,320	2.86	548	111	17.2	50.8	549	13.0
586	14.2	5.86	3,434	2.91	550	110	17.8	50.0	542	12.7
586	14.4	5.78	3,387	2.98	556	110	17.9	50.8	553	13.1
578	12.8	5.84	3,376	2.99	558	110	19.0	51.1	552	13.5
581	14.1	5.88	3,416	2.97	573	116	18.2	48.2	553	13.1
572	12.9	5.98	3,421	2.81	548	112	18.0	47.0	553	13.4
569	13.2	5.86	3,330	2.81	551	113	17.6	45.5	548	13.2
552	14.9	5.43	2,997	3.14	557	106	20.2	41.9	546	13.5
564	14.2	5.44	3,068	3.12	559	107	19.2	45.0	566	13.3
562	14.1	5.48	3,080	3.25	573	108	19.2	41.3	566	13.5
576	14.5	5.57	3,208	3.22	558	104	19.3	40.8	565	13.9
584	14.4	5.81	3,393	2.76	550	115	16.8	40.6	559	13.3
579	14.4	5.78	3,347	2.91	552	111	17.4	43.1	554	13.6
574	14.2	5.83	3,346	2.92	532	104	17.8	43.1	552	13.4
573	14.1	5.70	3,266	2.94	553	110	18.0	44.2	556	12.8
580	14.2	5.75	3,335	2.82	534	107	17.6	40.6	556	13.6
589	14.5	5.74	3,281	2.89	560	114	17.6	42.0	558	13.6
578	12.7	5.79	3,347	2.75	546	114	16.3	42.8	551	12.9
578	13.0	5.89	3,404	2.98	552	108	15.0	44.5	553	13.2
575	14.0	5.79	3,329	3.00	558	110	18.1	45.7	549	13.6
574	13.1	5.69	3,266	3.37	581	108	20.4	58.8	561	13.6
610	13.0	5.72	3,489	2.81	544	110	13.8	51.6	560	13.4
602	13.2	5.68	3,419	2.85	545	110	17.5	49.8	549	13.2
593	13.1	5.66	3,356	2.95	544	107	17.6	50.6	552	13.8
598	12.6	5.62	3,364	2.96	551	109	17.6	50.5	569	13.5

Brockton — Concluded.

COAL				WATER				COMMERCIAL		
B.T.U.	Candle Power.	Yield.	B.T.U. Feet.	Gals.	B.T.U.	Factor.	Candle Power.	Per Cent. H_2O.	B.T.U.	Candle Power.
558	12.4	5.54	3,258	3.29	588	112	20.3	51.6	574	13.2
601	12.3	5.51	3,312	2.85	558	114	17.2	53.2	571	13.5
584	12.8	5.51	3,218	3.05	549	106	18.3	56.4	569	13.6
602	12.8	5.35	3,221	3.24	559	104	19.8	66.1	577	13.7
543	12.8	5.25	2,851	2.97	595	124	18.4	76.8	580	13.4
474	12.7	4.67	2,214	3.08	578	114	18.8	77.9	557	13.5
564	13.0	5.36	3,023	3.24	605	118	19.5	70.9	575	14.3
548	12.9	5.34	2,926	2.83	557	115	17.1	73.0	561	14.1
583	12.1	5.48	3,195	2.98	560	111	17.9	71.8	558	14.4
584	12.9	5.43	3,171	3.14	575	111	18.7	78.1	570	13.8
582	13.1	5.52	3,213	3.19	580	112	19.2	76.2	576	13.9
566	13.5	5.42	3,068	3.01	551	104	18.1	80.0	555	12.8
578.5	13.8	5.67	3,289	2.98	560	112	17.7	51.2	559	13.8

Cambridge.

COAL				WATER				COMMERCIAL		
584	13.8	5.45	3,183	2.36	–	–	–	20.9	570	14.7
592	–	5.33	3,155	2.20	485	108	10.7	23.0	566	14.6
588	14.9	5.33	3,134	2.02	477	113	9.5	25.9	555	14.1
595	15.0	5.44	3,233	0.53	355	128	1.5	17.3	563	14.3
588	15.0	5.34	3,140	2.00	473	111	5.9	29.6	555	13.5
590	15.5	5.33	3,145	1.82	448	105	10.5	25.6	557	13.8
596	15.6	5.53	3,296	1.28	403	105	–	21.5	558	13.6
596	15.6	5.57	3,320	2.52	506	110	12.8	32.2	559	14.0
598	15.6	5.48	3,277	2.00	468	108	7.6	27.5	555	13.3
594	15.1	5.33	3,166	2.15	473	105	7.8	29.5	556	13.9
593	15.4	5.16	3,060	2.12	476	107	8.0	28.7	555	13.4
589	14.3	5.28	3,110	2.41	504	109	11.1	29.4	561	13.8
593	15.0	5.29	3,137	2.27	494	110	9.3	23.7	568	14.6
592	15.3	5.24	3,102	2.41	497	106	9.3	23.0	569	14.1
595	15.5	5.29	3,148	2.54	536	117	14.0	19.3	577	14.6
597	14.9	5.23	3,122	3.09	557	104	16.0	26.0	581	14.4
596	15.2	5.52	3,290	3.28	580	110	17.7	19.4	587	15.0
597	14.4	5.55	3,313	3.48	597	109	19.1	21.0	591	15.4
601	15.1	5.41	3,250	3.17	565	106	16.3	23.8	587	14.7

Cambridge — Continued.

COAL.				WATER.				COMMERCIAL.		
B.T.U.	Candle Power.	Yield.	B.T.U. Feet.	Gals.	B.T.U.	Factor.	Candle Power.	Per Cent. H₂O.	B.T.U.	Candle Power.
597	15.1	5.49	3,278	2.81	544	115	15.4	31.8	563	15.0
593	14.9	5.34	3,167	2.77	524	106	13.4	35.1	575	14.2
599	14.6	5.10	3,055	2.83	538	108	14.4	39.3	579	14.8
601	14.0	5.25	3,155	2.67	532	112	14.1	42.2	573	14.5
596	13.7	5.29	3,163	2.98	548	106	15.3	35.9	576	14.4
600	14.3	5.22	3,132	2.83	543	110	14.6	37.0	576	14.6
609	14.5	5.12	3,118	2.66	531	111	13.9	37.8	573	14.3
617	15.3	5.13	3,165	2.56	521	111	12.7	36.8	574	14.5
609	14.9	5.41	3,295	2.77	539	111	14.1	36.6	575	14.1
602	14.3	5.37	3,233	2.60	516	108	12.6	29.3	571	14.2
596	14.8	5.27	3,141	2.66	522	110	13.2	30.9	573	14.3
584	14.3	5.15	3,008	2.80	517	108	12.2	35.4	571	14.1
593	14.9	5.46	3,238	2.52	522	112	12.7	30.1	572	14.3
592	14.1	5.75	3,404	2.52	512	109	11.1	32.7	561	13.9
598	14.9	5.62	3,351	2.38	503	113	10.2	31.7	565	14.1
600	14.1	5.15	3,090	2.67	505	101	10.5	32.2	560	13.6
609	15.1	5.10	3,159	2.50	497	103	9.7	31.9	562	14.1
596	14.9	5.38	3,206	2.58	505	104	10.2	34.5	553	13.6
602	15.6	5.31	3,197	2.34	496	108	10.1	30.1	559	13.4
611	14.9	5.37	3,281	2.23	495	112	9.8	30.5	564	13.8
601	14.4	5.57	3,348	2.38	500	108	9.3	28.7	564	13.5
599	14.4	5.58	3,342	2.39	506	110	10.2	22.5	569	13.7
613	15.1	5.40	3,310	2.31	514	117	9.8	21.5	569	13.5
593	13.6	5.17	3,066	2.77	512	101	12.0	36.3	559	13.3
604	15.2	5.17	3,123	2.63	503	101	11.7	36.6	563	13.6
595	14.3	5.42	3,225	2.56	513	108	12.2	34.9	563	13.9
599	14.9	5.48	3,283	2.42	493	104	11.1	31.5	560	13.7
588	14.0	5.64	3,316	2.25	476	102	9.7	32.2	558	13.4
587	14.0	5.62	3,299	1.97	454	103	8.1	28.9	556	13.0
591	14.8	5.50	3,251	1.97	437	118	–	24.9	551	13.0
595	14.8	5.45	3,242	1.59	427	104	–	21.0	551	12.5
608	15.6	5.32	3,235	1.32	403	103	–	20.0	564	12.6
585	13.6	5.54	3,208	0.09	309	123	–	13.6	554	12.3
591	14.5	5.37	3,174	2.74	481	90	11.0	20.9	565	12.4
597	14.9	5.13	3,063	2.08	451	97	9.4	21.1	568	13.3

Cambridge — Concluded.

COAL.				WATER.				COMMERCIAL.		
B.T.U.	Candle Power.	Yield.	B.T.U. Feet.	Gals.	B.T.U.	Factor.	Candle Power.	Per Cent. H_2O.	B.T.U.	Candle Power.
600	15.4	5.18	3,108	1.86	438	98	6.7	23.3	553	13.0
597	14.8	5.33	3,182	2.10	461	100	7.7	27.2	553	13.6
583	14.2	5.43	3,166	2.44	488	101	9.4	28.7	554	13.1
578	13.4	5.30	3,063	2.61	518	108	12.2	28.7	548	12.9
586 ·	14.2	5.37	3,147	2.57	509	106	12.3	33.5	560	13.6
585	13.2	5.12	2,995	2.52	520	111	12.8	36.6	557	13.6
596	14.5	5.35	3,176	2.31	487	105	10.9	28.7	565	13.8

Charlestown.

B.T.U.	Candle Power.	Yield.	B.T.U. Feet.	Gals.	B.T.U.	Factor.	Candle Power.	Per Cent. H_2O.	B.T.U.	Candle Power.
611	16.8	5.26	3,215	-	-	-	-	-	-	-
612	16.7	5.30	3,243	-	-	-	-	-	-	-
617	16.6	5.27	3,251	-	-	-	-	-	-	-
607	16.5	5.40	3,297	-	-	-	-	-	-	-
617	16.7	5.45	3,352	-	-	-	-	-	-	-
630	16.7	5.58	3,514	-	-	-	-	-	-	-
625	17.0	5.38	3,365	-	-	-	-	-	-	-
620	16.9	5.22	3,236	-	-	-	-	-	-	-
630	16.8	5.38	3,393	-	-	-	-	-	-	-
615	16.4	5.25	3,228	-	-	-	-	-	-	-
617	16.6	5.38	3,319	-	-	-	- '	-	-	-
620	16.9	5.50	3,413	-	-	-	-	-	-	-
619	16.9	5.50	3,404	-	-	-	-	-	-	-
616	17.0	5.26	3,239	-	-	-	-	-	-	-
622	17.0	5.44	3,383	-	-	-	-	-	-	-
612	17.0	5.44	3,328	-	-	-	-	-	-	-
606	16.8	5.15	3,121	-	-	-	-	-	-	-
602	16.8	5.11	3,078	-	-	-	-	-	-	-
601	16.5	5.11	3,072	-	-	-	-	-	-	-
602	16.7	4.91	2,958	-	-	-	-	-	-	-
609	16.8	5.04	3,067	-	-	-	-	-	-	-
597	16.8	5.21	3,109	-	-	-	-	-	-	-
596	16.6	5.05	3,011	-	-	-	-	-	-	-
602	16.9	5.11	3,075	-	-	-	-	-	-	-
596	16.7	5.24	3,124	-	-	-	-	-	-	-

Charlestown — Concluded..

	COAL.				WATER.				COMMERCIAL.		
B.T.U.	Candle Power.	Yield.	B.T.U. Feet.	Gals.	B.T.U.	Factor.	Candle Power.	Per Cent. H₂O.	B.T.U.	Candle Power.
597	17.2	5.28	3,151	-	-	-	-	-	-	-
608	17.1	5.05	3,341	-	-	-	-	-	-	-
616	17.2	5.05	3,111	-	-	-	-	-	-	-
602	17.1	5.15	3,099	-	-	-	-	-	-	-
603	17.2	5.03	3,032	-	-	-	-	-	-	-
618	16.6	5.05	3,122	-	-	-	-	-	-	-
609	16.7	5.04	3,069	-	-	-	-	-	-	-
611	16.8	5.27	3,209	-	-	-	-	-	-	-

East Boston.

B.T.U.	Candle Power.	Yield.	B.T.U. Feet.	Gals.	B.T.U.	Factor.	Candle Power.	Per Cent. H₂O.	B.T.U.	Candle Power.
-	-	5.20	-	3.15	-	-	20.8	73.0	586	16.8
-	-	5.20	-	3.28	-	-	20.2	70.0	572	16.5
-	-	5.36	-	3.25	-	-	20.1	72.0	576	16.4
-	-	5.40	-	2.70	-	-	18.6	71.0	549	14.3
-	-	5.00	-	2.86	-	-	18.0	72.0	548	15.7
-	-	5.66	-	2.70	-	-	19.0	71.0	542	14.9
-	-	5.30	-	2.80	-	-	18.9	68.0	550	15.5
-	-	5.60	-	2.50	-	-	18.3	70.0	534	15.2
-	-	5.40	-	2.66	-	-	19.0	72.0	550	15.6
-	-	5.53	-	2.70	-	-	18.9	72.0	550	14.4
-	-	5.80	-	2.80	-	-	18.5	71.0	534	15.1
-	-	5.50	-	2.60	-	-	16.1	68.0	542	15.0
-	-	5.35	-	2.68	-	-	17.9	73.0	542	15.0
-	-	5.25	-	2.47	-	-	17.3	75.0	526	14.5
-	-	5.72	-	2.55	-	-	14.4	71.0	525	14.0
-	-	5.00	-	2.60	-	-	14.5	75.0	527	12.5
-	-	5.20	-	2.60	-	-	14.0	72.0	525	12.0
-	-	5.10	-	2.70	-	-	15.1	74.0	524	13.0
-	-	5.26	-	2.76	-	-	16.1	73.0	526	11.6
-	-	5.30	-	2.70	-	-	15.7	74.0	522	12.8
-	-	5.40	-	2.90	-	-	16.6	70.0	521	12.5
-	-	5.17	-	2.73	-	-	14.4	71.0	535	13.4
-	-	4.86	-	2.57	-	-	14.5	74.0	534	13.3
-	-	4.80	-	2.59	-	-	15.3	73.0	536	14.2

39

East Boston — Concluded.

COAL.				WATER.				COMMERCIAL.		
B.T.U.	Candle Power.	Yield.	B.T.U. Feet.	Gals.	B.T.U.	Factor.	Candle Power.	Per Cent. H₂O.	B.T.U.	Candle Power.
–	–	4.52	–	2.71	–	–	16.2	77.0	546	14.4
–	–	4.83	–	2.72	–	–	18.4	81.0	534	15.2
–	–	5.10	–	2.56	–	–	17.2	76.0	522	14.7
–	–	4.88	–	3.06	–	–	17.7	88.0	538	15.8
–	–	4.99	–	2.61	–	–	16.5	69.0	534	13.8
–	–	4.53	–	2.69	–	–	15.4	77.0	517	12.5
–	–	4.84	–	2.77	–	–	16.9	74.0	528	14.7
–	–	4.99	–	2.95	–	–	15.6	70.0	520	13.1
–	–	4.65	–	2.83	–	–	16.5	75.0	527	13.5
–	–	4.82	–	2.85	–	–	15.5	71.0	515	13.0
–	–	4.54	–	2.73	–	–	18.2	86.0	519	14.9
–	–	4.66	–	2.70	–	–	18.1	85.0	521	14.4
–	–	4.58	–	2.95	–	–	17.9	80.0	523	14.2
–	–	5.34	–	2.87	–	–	18.0	84.0	–	–
–	–	4.91	–	2.90	–	–	19.0	85.0	534	15.9
–	–	5.31	–	2.87	–	–	18.1	83.0	551	14.5
–	–	5.57	–	2.93	–	–	18.2	83.0	538	14.9
–	–	5.08	–	2.86	–	–	18.0	83.0	543	14.9
–	–	5.08	–	2.95	–	–	18.4	84.0	551	15.5
–	–	5.46	–	2.91	–	–	18.0	82.0	546	14.7
–	–	5.40	–	2.90	–	–	19.9	84.0	548	15.1
–	–	5.70	–	2.91	–	–	19.9	83.0	549	15.6
–	–	5.00	–	3.00	–	–	19.2	84.0	534	15.4
–	–	5.10	–	3.10	–	–	19.2	85.0	545	15.2
–	–	5.30	–	2.90	–	–	18.7	83.0	529	14.6
–	–	5.30	–	2.89	–	–	18.9	84.0	546	15.4
–	–	5.07	–	2.94	–	–	18.9	85.0	532	14.8
–	–	5.15	–	2.78	–	–	17.5	76.6	537	14.5

Fall River.

COAL.				WATER.				COMMERCIAL.		
B.T.U.	Candle Power.	Yield.	B.T.U. Feet.	Gals.	B.T.U.	Factor.	Candle Power.	Per Cent. H₂O.	B.T.U.	Candle Power.
565	15.0	5.67	3,200	2.81	545	111	19.1	–	548	15.3
551	14.2	5.63	3,100	2.88	556	113	18.6	–	554	15.2
573	15.1	5.65	3,240	2.89	540	107	18.7	–	563	16.7
564	15.0	5.66	3,190	2.86	534	106	18.4	–	565	–
–	14.9	5.71	–	2.94	533	103	18.5	47.0	549	14.5
553	15.9	5.72	3,163	2.99	558	110	18.4	47.3	563	15.4
580	16.3	5.69	3,295	2.86	541	108	17.3	48.0	562	15.7
567	15.8	5.59	3,165	3.00	546	106	17.3	51.0	553	15.3
553	15.1	5.84	3,220	3.06	558	108	17.7	50.7	550	15.1
543	14.9	5.67	3,075	3.02	558	108	17.8	51.6	561	–
560	15.4	5.83	3,265	2.89	533	101	17.9	50.7	556	14.9
561	16.3	5.68	3,185	2.96	556	110	18.3	48.1	560	15.5
564	15.7	5.87	3,310	3.11	548	102	18.3	48.4	562	15.1
568	15.3	5.62	3,190	3.00	550	107	18.2	48.2	563	14.6
567	15.7	5.66	3,205	3.05	553	107	18.1	47.3	556	14.2
566	15.7	5.67	3,210	3.16	–	–	18.0	46.6	562	14.2
551	15.2	5.80	3,195	3.08	549	105	17.8	44.1	555	14.2
560	15.7	5.67	3,175	2.95	550	109	17.7	48.0	556	14.1
559	16.0	5.82	3,255	2.88	–	–	17.5	46.0	556	14.1
563	15.5	5.52	3,108	3.01	549	107	17.9	48.9	547	13.7
560	15.6	5.72	3,145	3.12	–	–	17.6	50.5	553	14.2
558	15.4	5.79	3,235	2.92	–	–	17.8	48.7	554	14.8
560	16.4	5.60	3,135	2.92	–	–	17.8	48.9	557	15.4
573	16.3	5.35	3,065	2.96	–	–	17.7	50.4	559	14.9
576	15.6	5.46	3,145	3.02	–	–	17.6	50.8	552	14.2
566	15.8	5.58	3,160	3.03	–	–	17.3	50.5	560	14.3
563	15.5	5.59	3,145	3.04	–	–	17.2	45.5	553	13.9
574	15.7	5.07	2,910	3.11	–	–	17.5	53.4	561	14.1
577	15.7	5.01	2,890	2.98	–	–	17.1	56.2	561	14.4
569	15.9	5.30	3,015	3.10	–	–	17.2	46.4	563	14.7
569	16.3	5.50	3,130	3.00	–	–	16.9	46.7	562	14.8
570	16.1	5.52	3,145	3.04	–	–	16.9	44.0	560	14.6
551	14.9	5.84	3,215	3.03	–	–	16.9	42.5	546	13.7
543	14.7	5.78	3,140	2.97	–	–	16.6	42.5	548	13.7
541	14.8	5.58	3,020	2.96	–	–	16.8	43.2	552	13.9

Fall River — Concluded.

COAL				WATER				COMMERCIAL		
B.T.U.	Candle Power.	Yield.	B.T.U. Feet.	Gals.	B.T.U.	Factor.	Candle Power.	Per Cent. H_2O.	B.T.U.	Candle Power.
554	15.4	5.51	3,050	3.01	–	–	16.7	44.6	553	13.8
545	14.9	5.70	3,110	3.02	–	–	16.8	43.8	544	–
541	14.8	5.70	3,080	3.10	–	–	16.8	45.9	544	–
542	14.8	5.69	3,090	3.09	–	–	16.8	45.2	528	–
–	14.6	5.71	–	2.99	–	–	16.8	43.1	535	–
–	14.9	5.68	–	2.95	–	–	16.8	45.6	532	–
580	15.5	5.79	3,360	2.93	–	–	16.7	46.8	515	–
536	14.1	5.66	3,065	2.90	–	–	16.7	50.1	539	–
543	14.6	5.66	3,075	2.90	–	–	16.5	53.0	555	–
553	14.3	5.70	3,150	2.85	–	–	16.1	52.4	547	–
553	14.4	5.79	3,200	2.79	–	–	16.2	50.0	548	–
566	15.8	5.52	3,125	2.77	–	–	16.1	52.0	555	–
558	15.1	5.60	3,126	2.80	–	–	16.0	52.0	548	–
548	14.3	5.75	3,150	2.80	546	112	16.0	49.9	545	15.3
548	15.7	5.66	3,100	2.73	535	110	16.0	50.4	542	15.7
548	14.3	5.76	3,160	2.76	542	112	16.1	51.0	543	15.2
553	15.0	5.95	3,290	2.77	546	113	16.1	49.3	547	15.4
549	15.4	5.91	3,240	2.73	545	114	16.0	44.3	551	15.2
544	14.4	5.94	3,230	2.68	536	112	15.8	48.7	539	14.4
521	14.5	5.91	3,079	2.74	537	111	15.7	53.0	543	14.3
555	15.5	5.78	3,205	2.68	537	112	–	54.0	546	14.6
536	14.3	5.70	3,055	2.63	535	113	–	55.9	544	14.4
532	14.5	5.70	3,035	2.80	529	106	–	58.8	542	15.3
545	14.3	5.42	2,950	3.00	538	103	–	62.0	546	15.4
544	15.0	5.85	3,180	2.86	553	112	–	55.7	540	15.2
532	14.0	5.84	3,105	2.83	535	107	–	57.3	536	14.6
569	15.1	5.66	3,146	2.93	545	109	17.2	47.5	534	14.5

Fitchburg.

COAL				WATER				COMMERCIAL		
550	–	5.07	2,788	3.00	530	100	–	64.8	551	17.0
579	15.8	5.00	2,894	3.18	556	104	17.7	63.8	545	–
564	–	4.50	2,537	3.30	569	105	–	61.3	532	15.3
548	14.7	4.55	2,493	3.34	608	115	18.5	64.7	553	16.3
548	14.8	4.02	2,203	3.50	569	100	19.1	62.2	545	16.3

Fitchburg — Continued.

COAL				WATER				COMMERCIAL		
B.T.U.	Candle Power.	Yield.	B.T.U. Feet.	Gals.	B.T.U.	Factor.	Candle Power.	Per Cent. H$_2$O.	B.T.U.	Candle Power.
545	14.6	4.30	2,344	3.25	580	110	18.5	66.0	550	16.5
592	15.5	4.68	2,771	3.31	579	106	20.2	66.2	579	17.0
558	–	5.52	3,081	3.05	566	113	–	62.1	547	–
536	15.2	5.61	3,007	3.02	556	109	17.5	56.2	539	16.0
532	14.8	5.71	3,038	2.89	552	111	17.0	52.1	536	15.8
575	15.5	5.62	2,232	2.78	551	114	16.6	45.0	532	16.0
560	15.4	5.32	2,979	2.64	508	101	15.1	47.8	526	15.3
550	15.0	5.54	3,047	2.77	506	99	15.9	46.2	527	15.5
579	15.1	5.60	3,242	2.83	513	99	15.2	55.6	531	15.1
532	14.4	6.00	3,192	3.57	581	103	17.7	53.7	542	15.9
542	14.5	5.80	3,143	3.60	572	100	18.0	53.4	558	16.1
565	14.6	6.00	3,390	3.53	578	103	17.6	61.9	567	15.9
553	14.7	5.50	3,042	3.51	590	107	17.8	49.7	570	16.1
553	14.7	5.45	3,013	3.53	585	105	17.5	45.9	564	16.1
570	14.8	5.65	3,221	3.53	576	102	17.8	48.9	561	16.2
554	14.7	5.73	3,174	3.52	592	107	17.9	44.7	564	16.2
566	14.6	5.67	3,209	3.41	573	104	17.1	51.8	566	15.9
539	14.3	5.57	3,202	3.41	581	107	17.7	48.5	557	15.7
544	14.2	5.71	3,106	3.35	607	116	17.4	50.5	555	15.7
561	14.6	5.66	3,175	3.28	582	110	17.1	41.4	553	15.5
554	14.4	5.50	3,113	3.20	576	110	17.3	41.8	544	15.5
553	14.3	5.59	3,091	3.28	560	105	17.3	38.8	549	15.1
553	14.0	5.50	3,041	3.20	564	107	17.1	41.9	561	15.3
549	14.2	5.47	3,003	3.27	575	108	17.6	43.5	554	15.6
556	13.8	5.91	3,285	3.22	570	108	17.5	40.8	540	15.4
557	14.2	5.52	3,074	3.25	542	98	16.6	43.7	540	15.2
549	13.9	5.53	3,035	3.15	558	106	16.9	34.5	538	15.0
564	14.5	5.76	3,248	3.24	558	104	17.2	41.5	551	15.3
585	14.8	5.66	3,311	3.26	543	99	16.6	39.4	549	15.0
564	14.4	5.90	3,327	3.27	536	96	16.6	40.8	538	15.0
542	14.2	6.00	3,252	3.23	551	100	16.4	47.2	549	15.1
555	14.6	6.02	3,341	3.22	556	103	17.5	43.9	544	15.5
540	14.2	5.89	3,180	3.23	548	100	16.0	48.3	539	14.9
546	13.8	5.98	3,265	3.20	577	111	16.4	45.1	543	14.8
562	14.1	5.47	3,074	3.03	568	113	15.5	43.6	553	14.9

Fitchburg — Concluded.

	COAL.				WATER.				COMMERCIAL.	
B.T.U.	Candle Power.	Yield.	B.T.U. Feet.	Gals.	B.T.U.	Factor.	Candle Power.	Per Cent. H₂O.	B.T.U.	Candle Power.
539	13.6	5.93	3,196	2.91	553	111	15.3	49.1	542	14.4
551	13.8	5.92	3,261	2.92	548	109	15.6	44.4	543	14.4
559	14.6	5.67	3,169	2.84	542	109	15.1	46.1	554	14.7
561	14.7	5.55	3,113	2.68	533	111	14.8	48.6	540	14.7
567	14.4	5.77	3,271	2.65	534	112	14.0	54.3	538	14.1
529	12.9	5.94	3,284	2.61	533	113	15.4	54.0	531	13.1
528	12.5	6.00	3,120	2.99	550	108	16.1	61.0	535	13.7
559	12.8	6.00	3,354	3.06	552	106	16.3	59.0	540	14.3
523	12.2	6.30	3,295	2.80	541	110	15.4	58.0	529	13.6
549	13.4	6.10	3,370	2.96	539	105	15.4	61.0	534	14.7
535	13.7	5.90	3,167	2.92	537	105	15.5	54.0	543	15.2
550	14.2	5.84	3,212	2.93	537	105	15.8	53.0	543	15.0
549	12.8	5.49	3,014	2.91	520	100	15.7	56.0	545	14.0
552	13.4	5.49	3,030	2.78	533	108	16.0	57.0	543	14.8
574	14.6	5.53	3,174	2.86	535	106	15.4	58.0	550	15.2
539	13.2	5.59	3,013	2.88	528	103	14.6	68.0	530	13.5
521	13.0	5.84	3,022	3.01	545	105	16.2	62.0	538	14.8
526	12.9	5.48	2,882	3.22	540	98	16.8	63.0	535	–
529	13.0	5.48	2,898	3.25	555	102	16.6	67.0	543	15.1
521	13.0	5.47	2,850	3.89	572	96	20.2	39.0	549	15.1
552	14.1	5.58	3,064	3.14	557	106	16.7	52.1	545	15.2

Haverhill.

–	–	–	–	2.90	558	113	17.0	–	–	–
–	–	–	–	2.96	558	111	16.5	–	–	–
–	–	–	–	2.76	536	109	17.9	–	–	–
–	–	–	–	2.89	559	114	–	–	–	–
–	–	–	–	2.79	547	113	17.7	–	–	–
–	–	–	–	2.80	537	108	16.0	–	–	–
–	–	–	–	2.95	551	109	16.8	–	–	–
–	–	–	–	2.95	566	113	17.5	–	–	–
–	–	–	–	3.60	558	96	16.4	–	–	–
–	–	–	–	3.09	570	111	17.7	–	–	–
–	–	–	–	2.90	559	113	16.8	–	–	–

Haverhill — Continued.

	COAL.				WATER.				COMMERCIAL.	
B.T.U.	Candle Power.	Yield.	B.T.U. Feet.	Gals.	B.T.U.	Factor.	Candle Power.	Per Cent. H₂O.	B.T.U.	Candle Power.
-	-	-	-	3.08	566	110	16.7	-	-	-
-	-	-	-	3.03	557	109	19.3	-	-	-
-	⌐	-	-	3.13	555	104	18.5	-	-	-
-	-	-	-	3.06	554	107	18.9	-	-	-
-	-	-	-	3.05	553	107	19.1	-	-	-
-	-	-	-	3.03	549	106	19.7	-	-	-
-	-	-	-	3.04	546	105	16.1	-	-	-
-	-	-	-	2.96	552	109	18.6	-	-	-
-	-	-	-	2.94	555	107	17.6	-	-	-
-	-	-	-	2.92	544	108	18.3	-	565	17.3
-	-	-	-	-	554	-	17.8	-	561	16.9
-	-	-	-	3.17	552	101	17.4	-	550	16.1
-	-	-	-	2.96	546	107	16.6	-	546	15.9
-	-	-	-	3.15	552	101	17.5	-	552	16.0
-	-	-	-	2.97	549	108	17.5	-	560	15.9
-	-	-	-	2.94	544	107	16.4	-	550	15.7
-	-	-	-	2.82	539	109	15.5	-	551	15.6
-	-	-	-	2.81	544	110	16.4	-	551	15.6
-	-	-	-	2.85	542	112	15.2	-	542	14.7
-	-	-	-	2.88	547	110	15.8	-	547	15.0
-	-	-	-	2.92	555	111	16.0	-	-	-
-	-	-	-	2.81	560	116	15.7	-	560	15.8
-	-	-	-	2.84	545	110	14.3	-	545	14.9
-	-	-	-	2.84	553	113	15.0	-	553	15.5
-	-	-	-	2.73	549	115	15.5	-	549	15.0
-	-	-	-	2.87	552	109	15.3	-	552	15.4
-	-	-	-	2.99	552	108	15.6	-	552	16.1
-	-	-	-	2.78	551	114	14.9	-	551	16.1
-	-	-	-	2.73	556	118	15.2	-	556	16.0
-	-	-	-	2.82	548	112	14.2	-	548	15.6
-	-	-	-	2.86	553	113	14.6	-	553	16.3
-	-	-	-	2.82	551	113	14.3	-	551	16.1
-	-	-	-	2.69	547	116	15.4	-	547	15.3
-	-	-	-	2.71	554	118	16.7	-	554	15.7
-	-	-	-	2.64	542	116	15.7	-	542	15.1

Haverhill — Concluded.

COAL.				WATER.				COMMERCIAL.		
B.T.U.	Candle Power.	Yield.	B.T.U. Feet.	Gals.	B.T.U.	Factor.	Candle Power.	Per Cent. H₂O.	B.T.U.	Candle Power.
-	-	-	-	2.63	534	113	13.4	-	534	14.6
-	-	-	-	2.73	543	113	14.7	-	543	14.9
-	-	-	-	2.47	532	118	13.9	-	532	14.4
-	-	-	-	2.75	535	110	14.2	-	535	14.2
-	-	-	-	2.64	538	114	14.6	-	538	14.5
-	-	-	-	2.63	522	108	13.3	-	522	13.8
-	-	-	-	2.72	524	107	14.2	-	524	14.1
-	-	-	-	2.88	537	105	15.6	-	537	14.6
-	-	-	-	2.72	529	108	14.6	-	529	14.3
-	-	-	-	2.93	536	104	14.8	-	536	14.6
-	-	-	-	2.76	-	-	-	-	-	15.5
-	-	-	-	2.63	530	111	15.7	-	530	14.7
-	-	-	-	2.56	525	112	14.4	-	525	14.8
-	-	-	-	2.87	547	110	16.1	-	545	15.3

Holyoke.

COAL.				WATER.				COMMERCIAL.		
B.T.U.	Candle Power.	Yield.	B.T.U. Feet.	Gals.	B.T.U.	Factor.	Candle Power.	Per Cent. H₂O.	B.T.U.	Candle Power.
592	14.4	5.23	3,096	2.84	550	112	16.1	45.0	571	14.8
593	15.7	5.17	3,066	3.05	569	112	14.0	48.0	579	15 2
-	-	-	-	3.82	650	116	19.1	-	-	-
603	16.3	-	-	-	574	-	16.0	75.0	581	16.0
590	15.5	-	-	-	626	-	17.4	75.0	610	16.7
581	15.3	5.35	3,108	3.39	568	103	17.1	35.0	573	-
-	-	5.69	-	3.20	-	-	-	43.0	-	-
588	14.4	5.36	3,152	3.60	592	105	17.1	56.0	581	-
590	15.2	5.42	3,198	3.42	583	107	17.4	30.0	-	-
599	15.0	5.13	3,073	3.39	603	113	16.9	36.0	601	15.7
585	14.1	5.46	3,194	3.57	619	114	17.6	39.0	-	-
582	14.7	5.08	2,957	3.39	602	112	16.7	53.0	-	-
570	13.4	5.18	2,953	3.24	590	116	16.0	46.0	546	14.3
570	13.0	4.78	2,725	3.35	577	107	16.2	53.0	558	13.3
-	-	4.83	-	3.36	-	-	-	55.0	548	-
572	12.0	4.25	2,431	3.15	554	105	15.8	63.0	545	-
575	13.0	4.52	2,609	3.19	560	106	15.7	59.0	537	-
573	13.2	4.73	2,710	3.10	570	111	14.7	68.0	555	-

Holyoke — Concluded.

COAL.				WATER.				COMMERCIAL.		
B.T.U.	Candle Power.	Yield.	B.T.U. Feet.	Gals.	B.T.U.	Factor.	Candle Power.	Per Cent. H₂O.	B.T.U.	Candle Power.
575	14.2	4.30	2,473	3.06	563	110	15.4	71.0	552	–
560	13.1	4.50	2,520	3.12	582	115	15.8	70.0	559	–
570	14.2	4.47	2,548	2.76	557	117	13.7	65.0	546	–
579	14.0	4.73	2,739	2.72	571	124	13.3	61.0	557	–
573	13.6	4.74	2,716	2.72	529	108	14.3	67.0	527	–
553	13.1	4.04	2,240	2.98	561	112	15.3	71.0	544	–
587	15.4	4.23	2,483	3.02	581	117	15.5	80.0	469	–
555	14.2	5.55	3,080	2.93	563	116	15.0	66.0	546	–
581	14.7	5.20	3,021	3.62	600	107	18.6	63.0	577	–
–	–	5.30	–	3.51	–	–	–	65.0	562	–
575	14.2	5.62	3,232	3.28	587	112	15.9	75.0	660	–
563	13.3	5.44	3,062	3.14	583	114	15.6	73.0	589	–
562	12.3	5.30	2,979	2.95	534	103	15.1	66.0	541	–
583	13.0	5.39	3,142	2.90	540	107	14.3	75.0	555	–
563	13.6	4.67	2,629	2.86	544	109	15.2	80.0	541	–
576	14.1	4.80	2,765	2.97	551	109	15.1	80.0	562	–
572	14.7	4.77	2,728	2.90	557	113	15.2	72.0	547	–
569	14.3	5.22	2,970	2.97	564	113	14.6	80.0	548	–
571	10.3	5.33	3,043	3.05	580	115	15.0	76.0	554	–
585	12.2	4.77	2,790	3.09	586	116	15.4	79.0	571	–
573	11.2	5.18	2,968	2.93	597	126	15.1	79.0	577	15.9
575	7.7	3.50	2,012	3.13	575	111	14.2	86.0	553	–
565	8.7	4.31	2,435	2.92	590	123	14.7	80.0	581	–
581	9.2	4.35	2,527	3.00	580	117	15.0	91.0	–	–
577	13.4	4.92	2,826	3.16	576	111	15.8	64.6	561	15.2

Lowell.

COAL.				WATER.				COMMERCIAL.		
593	14.8	5.30	3,143	3.78	578	98	18.0	45.0	590	16.6
584	16.1	5.30	3,095	3.94	573	93	17.8	17.5	584	16.6
596	14.3	5.38	3,205	3.45	570	103	16.7	22.9	591	15.0
595	15.4	5.37	3,195	3.52	569	100	17.2	21.4	583	15.7
594	14.2	5.40	3,207	3.50	567	100	16.7	25.9	575	15.5
581	13.8	5.42	3,150	3.76	592	102	18.1	30.4	584	14.8
598	14.6	5.38	3,216	3.41	569	103	17.6	33.0	588	14.9

Lowell — Continued.

COAL.				WATER.				COMMERCIAL.		
B.T.U.	Candle Power.	Yield.	B.T.U. Feet.	Gals.	B.T.U.	Factor.	Candle Power.	Per Cent. H₂O.	B.T.U.	Candle Power.
591	14.7	5.40	3,192	3.33	559	100	16.6	33.5	583	14.8
591	15.1	5.38	3,180	3.45	572	103	17.2	32.0	592	15.5
600	15.8	5.39	3,232	3.79	604	104	19.3	28.3	597	15.4
599	15.3	5.38	3,229	4.03	620	103	19.0	29.3	604	16.7
599	15.2	5.36	3,213	3.78	601	104	18.9	31.2	599	15.9
593	15.0	5.41	3,207	3.66	601	106	19.3	31.8	595	16.1
594	15.0	5.40	3,209	3.73	600	104	18.8	31.2	596	16.1
596	14.9	5.39	3,215	3.80	601	103	18.8	27.8	600	15.5
595	14.5	5.37	3,194	3.85	591	100	18.2	34.2	601	15.6
608	15.2	5.39	3,183	3.79	600	103	18.9	33.6	604	15.7
591	15.4	5.43	3,206	3.58	596	107	18.8	31.6	592	15.2
584	14.7	5.45	3,180	3.55	579	103	18.2	31.5	584	15.3
582	13.3	5.39	3,139	3.33	558	97	16.3	35.2	583	14.7
586	13.9	5.44	3,185	3.24	551	101	15.4	34.7	572	14.8
586	14.1	5.42	3,177	3.33	550	99	15.6	33.4	579	13.9
612	16.0	5.39	3,299	3.20	559	105	16.6	35.8	585	14.3
602	15.0	5.38	3,242	3.10	551	105	15.5	35.3	589	14.5
597	13.2	5.37	3,206	3.24	542	99	14.1	34.7	588	14.0
604	15.1	5.36	3,238	3.22	564	106	15.2	34.6	590	14.4
606	14.8	5.34	3,238	2.90	541	107	14.5	36.2	577	14.3
611	14.9	5.36	3,275	2.89	534	105	14.1	34.0	577	13.8
600	13.4	5.43	3,258	3.01	524	98	12.5	32.1	572	13.3
590	14.9	5.37	3,171	2.55	507	107	12.7	29.0	577	13.3
603	14.8	5.33	3,215	2.40	497	106	12.5	31.9	573	13.5
591	14.7	5.32	3,142	2.36	481	101	10.9	28.1	567	13.9
599	15.1	5.27	3,154	2.44	484	98	10.4	28.6	568	14.3
593	15.3	5.32	3,153	2.26	479	103	9.8	27.5	568	13.5
591	14.9	5.37	3,174	2.17	451	94	10.1	26.7	562	13.7
600	15.6	5.42	3,252	2.26	465	97	10.0	24.2	569	13.0
594	15.0	5.42	3,219	2.22	470	100	9.1	25.4	568	12.5
607	15.4	5.38	3,266	2.27	479	103	9.5	23.3	586	13.2
600	15.1	5.39	3,234	2.12	466	103	9.3	25.3	585	13.3
598	15.6	5.37	3,119	2.22	463	97	9.1	24.7	587	12.9
607	15.5	5.30	3,417	2.16	460	98	8.9	27.7	579	12.5
607	15.0	5.32	3,213	2.17	462	99	8.6	25.3	580	12.3

Lowell — Concluded.

COAL.				WATER.				COMMERCIAL.		
B.T.U.	Candle Power.	Yield.	B.T.U. Feet.	Gals.	B.T.U.	Factor.	Candle Power.	Per Cent. H₂O.	B.T.U.	Candle Power.
663	14.7	5.35	3,126	2.21	465	94	9.1	27.2	574	12.5
584	14.2	5.40	3,154	2.61	475	91	10.1	36.6	577	11.2
593	15.5	5.36	3,205	2.48	490	100	11.1	27.5	586	12.5
594	15.3	5.28	3,136	2.57	482	95	10.8	28.5	586	13.1
605	15.7	5.21	3,152	3.38	506	114	12.6	27.6	588	13.1
604	15.8	5.31	3,207	3.39	531	92	14.0	28.6	584	14.3
608	15.5	5.37	3,265	3.01	544	105	14.5	26.9	587	14.7
596	14.9	5.41	3,224	2.97	513	96	13.0	27.3	584	14.2
599	16.1	5.38	3,222	3.01	533	101	13.5	26.9	587	15.1
591	15.7	5.41	3,215	3.13	533	98	14.6	24.2	586	15.6
607	16.5	5.65	3,480	3.22	545	100	15.8	29.9	590	15.7
597	15.1	5.39	3,218	3.03	540	103	14.0	29.5	589	14.6
573	14.9	5.38	3,083	3.10	509	91	13.1	26.5	562	14.2
–	15.5	5.37	–	3.19	–	–	13.7	31.3	–	14.2
–	15.5	5.35	–	3.07	–	–	13.3	27.8	–	14.1
–	15.0	5.46	–	3.12	–	–	13.4	30.3	–	13.3
582	15.3	5.42	3,154	3.37	526	92	12.7	28.3	573	14.8
591	16.3	5.32	3,150	3.63	557	95	17.0	28.1	591	15.6
572	14.9	5.42	3,100	3.94	561	90	15.3	31.2	576	14.4
594	15.0	5.43	3,201	3.08	536	100	14.3	29.6	584	14.3

Lynn.

COAL.				WATER.				COMMERCIAL.		
B.T.U.	Candle Power.	Yield.	B.T.U. Feet.	Gals.	B.T.U.	Factor.	Candle Power.	Per Cent. H₂O.	B.T.U.	Candle Power.
577	14.7	5.33	3,075	2.99	567	113	19.0	50.0	574	16.5
568	13.3	5.37	3,050	3.03	579	116	19.5	–	580	16.8
564	15.3	5.25	2,961	2.97	575	117	18.7	–	581	17.8
617	14.7	5.20	3,208	3.06	573	113	17.8	–	–	–
578	13.8	5.25	3,034	–	–	–	–	–	578	15.8
574	16.2	5.18	2,973	2.88	548	110	17.1	–	579	16.7
541	15.1	5.08	2,748	2.84	543	109	16.9	–	565	16.1
596	15.5	5.13	3,057	2.94	544	105	16.8	–	566	16.3
552	15.2	5.47	–	2.87	517	103	–	50.0	560	16.5
567	–	5.37	3,044	2.74	509	104	–	38.0	547	–
571	–	5.24	2,992	2.68	518	110	–	41.0	549	16.9
562	–	5.28	2,970	2.66	519	111	–	38.0	556	–

Lynn — Continued.

COAL				WATER				COMMERCIAL		
B.T.U.	Candle Power.	Yield.	B.T.U. Feet.	Gals.	B.T.U.	Factor.	Candle Power.	Per Cent. H₂O.	B.T.U.	Candle Power.
577	–	5.15	2,961	2.73	518	107	–	–	549	14.1
587	15.3	5.15	3,023	2.95	548	108	17.4	45.0	566	15.5
581	15.9	5.15	2,992	3.06	571	113	17.0	50.0	562	–
590	14.7	5.08	2,997	3.15	565	108	16.9	49.0	570	15.1
595	14.8	5.35	3,183	2.96	547	107	17.6	48.0	564	–
569	–	5.35	3,044	2.96	555	110	–	48.0	562	–
565	–	5.43	3,068	3.01	565	112	–	58.7	555	–
549	–	5.32	2,920	2.95	573	116	–	62.8	557	–
559	–	5.29	2,957	2.79	558	117	–	58.0	560	–
590	–	5.26	3,103	2.85	554	113	–	59.0	555	–
577	–	5.09	2,937	3.07	550	106	–	64.0	553	–
576	–	5.52	3,179	3.04	545	105	–	59.0	550	–
571	–	5.50	3,140	3.14	551	104	–	55.0	552	–
565	–	5.51	3,113	3.41	558	100	–	56.3	560	–
570	–	5.46	3,112	3.21	570	108	–	57.0	556	–
567	–	5.58	3,163	3.07	561	109	–	51.2	550	–
565	–	5.45	3,079	3.16	567	109	–	46.8	549	–
550	14.0	5.43	2,987	3.11	551	105	–	45.0	550	13.8
551	11.8	5.24	2,887	3.28	563	104	–	43.7	562	14.6
571	12.6	5.16	2,946	3.49	570	101	–	38.8	568	14.0
570	11.3	5.19	2,958	3.25	558	103	17.7	39.0	573	14.4
580	12.9	5.35	3,103	2.84	530	105	17.7	33.0	562	14.4
–	–	5.31	–	2.81	534	106	–	34.3	554	13.5
561	12.1	5.19	2,911	2.84	526	104	12.9	34.5	557	13.5
554	10.7	5.29	2,930	2.73	521	105	14.0	33.2	555	13.6
–	–	5.23	–	2.74	530	108	–	35.9	556	13.6
–	–	5.21	–	2.66	529	110	–	39.3	552	13.6
583	13.1	5.23	3,049	2.51	509	107	14.3	37.3	555	13.9
588	11.1	5.12	3,010	2.50	500	104	11.5	35.5	563	13.6
–	–	5.34	–	2.51	509	107	–	42.7	551	12.8
564	–	5.26	2,966	2.47	–	–	–	43.0	537	13.3
553	–	5.35	2,958	2.52	503	105	–	43.6	537	12.4
561	–	5.30	2,973	2.48	495	103	–	43.0	536	12.2
550	–	5.05	2,770	2.54	495	100	–	46.0	540	12.6
565	–	4.94	2,791	2.57	502	103	–	51.0	542	12.4

Lynn — Concluded.

COAL.				WATER.				COMMERCIAL.		
B.T.U.	Candle Power.	Yield.	B.T.U. Feet.	Gals.	B.T.U.	Factor.	Candle Power.	Per Cent. H₂O.	B.T.U.	Candle Power.
567	–	5.12	2,903	2.56	495	100	–	47.9	535	–
574	–	5.02	2,881	2.47	485	99	–	49.6	529	12.7
571	–	5.02	2,866	2.51	494	101	–	45.5	531	12.8
510	–	5.06	2,580	2.47	515	111	–	45.4	535	13.0
526	–	4.95	2,603	2.48	488	100	–	49.4	530	12.6
547	–	4.80	2,625	2.48	492	105	–	53.6	528	12.7
548	–	4.92	2,697	2.47	508	109	–	51.2	526	12.5
569	–	4.97	2,817	2.74	508	100	–	51.4	539	–
590	–	5.14	3,032	2.99	550	108	–	51.4	544	–
573	–	5.23	2,996	3.02	549	107	–	49.0	546	13.1
560	–	5.12	2,867	2.91	531	100	–	46.8	546	13.1
553	–	5.08	2,809	2.94	540	105	–	48.1	546	13.1
562	–	4.98	2,798	2.96	530	110	–	46.3	546	13.4
567	13.8	5.22	2,962	2.85	535	107	16.2	46.9	553	14.1

Malden.

B.T.U.	Candle Power.	Yield.	B.T.U. Feet.	Gals.	B.T.U.	Factor.	Candle Power.	Per Cent. H₂O.	B.T.U.	Candle Power.
575	14.9	4.58	2,633	3.14	575	111	17.2	70.5	566	15.8
585	14.9	4.83	2,835	3.28	522	92	16.3	67.0	557	15.6
575	14.9	4.87	2,790	3.07	548	105	16.2	68.6	560	15.7
578	14.9	4.83	2,792	3.02	561	111	16.2	70.9	564	15.7
585	15.1	4.88	2,854	3.10	561	108	16.2	71.2	559	15.1
563	14.6	4.90	2,759	2.88	546	109	15.9	69.8	544	14.9
574	14.6	4.85	2,784	2.55	520	114	14.9	70.7	533	14.5
562	14.4	4.88	2,743	2.78	517	102	14.7	70.1	522	13.6
549	13.9	4.88	2,677	3.02	533	101	16.0	69.0	545	14.1
557	14.1	4.85	2,701	2.76	536	108	15.7	70.2	533	14.0
567	13.9	4.90	2,778	3.08	546	104	15.7	66.0	551	14.3
570	14.3	4.90	2,793	3.05	548	105	15.3	66.0	546	14.0
573	14.1	4.81	2,756	3.00	543	105	15.9	67.0	558	14.1
565	14.2	4.62	2,610	2.97	551	108	16.2	69.9	547	14.7
550	14.1	4.70	2,585	3.01	556	109	15.7	68.0	538	14.1
566	14.2	4.69	2,655	2.95	541	106	15.7	70.0	541	14.0
570	14.3	4.75	2,708	3.03	529	100	15.5	68.0	539	13.9
558	13.9	4.70	2,623	3.08	543	103	15.8	70.0	542	13.9

Malden — Continued.

	COAL.				WATER.				COMMERCIAL.	
B.T.U.	Candle Power.	Yield.	B.T.U. Feet.	Gals.	B.T.U.	Factor.	Candle Power.	Per Cent. H_2O.	B.T.U.	Candle Power.
561	13.9	4.59	2,575	3.01	547	106	15.8	71.7	542	14.0
560	13.7	4.73	2,648	3.15	538	100	15.7	70.0	543	13.8
558	13.9	4.81	2,684	3.26	543	99	15.9	69.0	543	14.1
556	13.7	4.90	2,729	3.13	556	106	16.8	71.0	548	14.3
559	13.5	4.76	2,661	2.78	557	117	16.8	74.0	548	14.2
556	12.9	4.70	2,613	3.19	564	107	17.1	75.8	551	14.4
548	12.1	4.85	2,658	3.15	546	102	16.2	74.9	526	13.2
540	11.5	4.84	2,614	3.12	558	107	16.7	75.0	534	13.4
555	12.1	4.74	2,631	3.10	560	108	17.3	75.0	550	14.4
567	13.2	4.66	2,642	3.18	565	109	17.3	73.8	554	15.1
548	11.9	4.74	2,598	2.97	545	106	16.6	72.6	531	13.8
554	12.6	4.79	2,654	3.01	547	106	16.9	72.9	535	13.4
567	12.8	4.84	2,744	2.98	545	106	16.7	72.5	536	13.4
573	13.1	4.81	2,756	2.93	542	103	16.2	67.3	543	13.7
566	12.6	4.90	2,773	2.86	534	106	15.8	63.0	535	13.5
567	12.2	4.90	2,778	2.92	529	103	16.3	63.0	531	13.1
560	12.2	4.90	2,744	2.92	513	97	15.7	62.0	523	12.3
567	12.8	4.90	2,778	2.82	511	95	15.9	58.9	531	12.9
586	12.9	4.69	2,748	2.86	524	103	16.3	67.0	542	13.3
585	13.5	4.76	2,785	2.94	521	99	15.8	66.8	544	13.7
548	11.2	4.89	2,680	2.93	505	94	15.8	65.8	520	12.4
568	12.1	4.84	2,749	2.85	528	104	16.6	67.1	543	13.5
565	12.1	4.56	2,576	2.97	552	106	16.5	70.1	549	14.6
558	12.0	4.65	2,595	3.01	544	105	16.9	70.3	541	13.9
563	11.7	4.83	2,719	3.10	550	105	17.1	70.3	544	14.1
565	11.5	4.77	2,735	3.04	534	101	16.2	70.7	542	14.0
564	11.6	4.63	2,611	2.95	540	105	16.5	73.0	545	14.2
556	11.1	4.58	2,537	2.94	551	110	16.9	75.0	542	14.0
534	9.9	4.66	2,488	2.98	542	105	16.5	75.0	530	12.7
533	10.0	4.68	2,494	2.89	539	107	16.7	75.0	533	13.9
528	10.0	4.53	2,392	2.86	543	109	17.4	75.9	534	13.6
528	10.1	4.89	2,582	2.86	541	108	17.1	73.8	535	13.8
535	9.9	4.90	2,622	2.99	539	104	17.1	74.4	534	14.4
539	11.5	4.90	2,641	2.88	536	106	16.9	75.8	535	13.9
547	10.0	4.90	2,680	2.87	548	111	18.2	75.0	539	14.3

Malden — Concluded.

COAL.				WATER.				COMMERCIAL.		
B.T.U.	Candle Power.	Yield.	B.T.U. Feet.	Gals.	B.T.U.	Factor.	Candle Power.	Per Cent. H₂O.	B.T.U.	Candle Power.
550	10.3	4.90	2,695	2.93	544	107	17.2	76.9	538	14.5
538	9.8	4.90	2,636	2.95	537	108	16.6	81.4	531	14.0
519	9.1	4.90	2,543	2.92	546	108	17.2	81.6	530	14.1
521	8.7	4.90	2,553	2.93	547	108	17.0	84.4	535	-
505	9.2	4.90	2,475	2.90	547	110	17.2	81.5	535	14.3
510	8.7	4.90	2,499	2.89	546	109	17.0	79.3	528	13.9
557	12.4	4.79	2,666	2.97	542	106	16.4	71.3	541	14.0

New Bedford.

B.T.U.	Candle Power.	Yield.	B.T.U. Feet.	Gals.	B.T.U.	Factor.	Candle Power.	Per Cent. H₂O.	B.T.U.	Candle Power.
623	15.3	5.07	3,159	3.48	562	99	15.1	70.0	560	15.2
627	15.3	5.17	3,242	3.38	609	115	15.1	70.0	550	15.3
627	15.3	5.15	3,229	3.59	582	103	17.0	66.0	578	16.8
626	15.4	5.12	3,205	2.64	531	111	14.0	70.0	516	14.8
623	15.1	5.12	3,190	3.79	595	100	17.6	72.0	577	16.9
627	14.9	4.97	3,116	3.44	592	109	17.6	69.0	591	16.8
619	14.8	5.16	3,194	3.35	579	107	17.3	75.0	570	16.6
598	14.5	4.98	2,969	3.36	554	100	17.1	78.0	559	16.7
631	14.7	5.16	3,256	3.24	565	106	17.4	76.0	567	16.7
623	14.4	5.24	3,265	3.13	561	107	17.3	78.0	572	16.6
617	14.7	5.05	3,116	3.20	560	105	17.5	78.0	573	16.5
622	14.7	5.02	3,122	3.20	572	109	17.3	76.0	575	16.4
-	14.2	5.11	-	3.27	-	-	17.6	74.0	569	16.3
-	12.3	5.23	-	3.38	-	-	16.9	73.0	554	16.0
-	12.6	5.10	-	2.90	-	-	16.3	75.0	547	15.5
-	12.6	5.21	-	3.11	-	-	16.7	73.0	556	15.5
604	14.0	5.14	3,105	3.16	547	96	17.1	72.0	570	16.1
608	14.5	5.25	3,192	3.10	557	107	18.2	73.0	570	17.1
619	14.7	5.20	3,219	3.63	583	102	19.6	72.0	590	18.1
613	13.5	5.10	3,126	3.41	575	105	17.7	74.0	582	17.0
622	13.9	5.05	3,141	3.26	570	107	18.4	74.0	576	16.9
613	14.1	5.20	3,188	3.42	573	103	18.4	72.0	578	17.1
614	14.2	5.05	3,101	3.11	575	112	18.0	73.0	574	16.7
611	14.0	5.11	3,122	2.93	544	108	17.1	75.0	552	16.2
609	14.0	5.21	3,173	2.92	573	118	16.9	74.0	558	16.2

New Bedford — Concluded.

COAL.				WATER.				COMMERCIAL.		
B.T.U.	Candle Power.	Yield.	B.T.U. Feet.	Gals.	B.T.U.	Factor.	Candle Power.	Per Cent. H₂O.	B.T.U.	Candle Power.
613	16.3	5.20	3,188	3.00	558	110	17.2	73.0	560	16.2
611	14.1	5.30	3,238	3.04	552	107	17.6	72.0	564	16.3
612	14.0	5.23	3,201	3.30	560	103	17.1	72.0	554	16.2
608	13.5	5.24	3,186	3.22	551	102	16.5	72.0	563	15.6
612	13.6	5.05	3,091	3.22	560	105	17.0	71.0	573	15.8
604	14.2	4.81	2,905	3.04	557	109	16.7	73.0	578	16.0
610	13.8	4.97	3,032	2.95	557	111	16.7	77.0	568	16.1
608	14.1	4.97	3,022	2.93	559	112	17.1	76.0	565	16.0
616	14.7	4.80	2,956	2.94	561	113	17.2	75.0	567	16.2
615	14.8	4.98	3,063	2.70	543	114	17.0	74.0	553	15.9
611	14.4	5.03	3,073	2.98	544	106	16.9	74.0	552	16.0
606	14.7	4.86	2,945	2.84	544	110	16.9	77.0	544	15.8
590	14.0	5.14	3,032	3.14	536	99	17.4	75.0	538	15.9
587	14.7	4.98	2,923	2.90	538	106	16.8	75.0	533	15.8
590	14.6	4.92	2,903	2.90	535	105	16.5	79.0	535	15.6
601	14.5	5.01	3,011	3.00	545	106	16.8	76.0	543	15.8
598	14.8	4.84	2,894	2.92	534	104	16.7	77.0	540	15.9
596	14.1	4.94	2,944	2.86	530	105	16.7	77.0	535	15.6
613	15.1	4.80	2,942	3.00	520	97	16.4	73.0	535	15.6
604	14.6	4.99	3,014	3.10	544	103	17.3	75.0	544	15.9
611	14.2	4.94	3,018	2.97	548	108	17.0	74.0	552	15.9
610	15.0	4.65	2,836	2.84	523	102	16.7	78.0	537	15.6
615	14.1	4.62	2,841	2.97	527	100	17.5	78.0	554	16.4
614	14.1	4.63	2,843	2.97	536	104	17.0	76.0	570	16.8
596	14.2	4.65	2,861	2.85	550	112	-	76.0	575	17.7
582	13.6	4.76	2,770	2.97	521	98	-	76.5	557	16.8
599	15.3	4.43	2,654	2.92	547	109	18.7	80.0	568	17.6
586	13.2	4.02	2,356	2.77	552	115	16.6	86.0	553	15.5
585	12.9	4.63	2,709	3.05	552	107	17.4	83.0	582	17.1
568	11.4	4.73	2,687	3.12	512	92	12.8	78.7	571	16.0
585	12.2	4.57	2,673	3.19	578	111	17.3	81.0	550	14.4
576	12.4	4.66	2,684	3.04	511	93	14.2	81.6	568	14.5
598	13.9	4.32	2,583	3.10	524	91	15.7	83.3	532	13.0
600	14.4	4.69	2,760	3.18	517	92	11.1	81.0	522	12.0
597	14.0	4.81	2,872	3.22	547	101	13.9	82.1	551	13.7
607	14.1	4.96	3,002	3.12	552	105	16.8	74.1	559	16.3

Old Colony.

WATER.			WATER.			WATER.		
Gallons.	B. T. U.	Factor.	Gallons.	B. T. U.	Factor.	Gallons.	B. T. U.	Factor.
2.82	557	115	2.80	528	105	2.50	514	110
2.90	544	108	2.80	524	104	2.70	526	108
2.70	528	110	2.90	528	104	2.40	519	116
2.60	532	113	2.50	524	114	2.40	522	117
2.66	530	110	2.40	527	117	2.40	510	111
2.80	528	105	2.80	528	106	2.50	516	110
2.70	530	113	2.90	533	105	2.50	513	109
2.60	523	109	2.80	534	108	2.50	514	110
2.70	523	110	2.80	528	104	2.30	514	117
2.60	522	109	2.80	527	105	2.40	520	116
2.30	509	115	2.90	526	100	2.50	515	110
2.70	510	102	2.80	528	105	3.00	500	91
2.90	517	99	2.60	533	102	2.50	512	109
2.90	544	108	2.70	532	110	2.50	510	108
2.80	536	108	2.60	530	112	2.60	514	107
2.30	524	121	2.70	532	110	2.50	507	107
2.80	530	106	2.50	531	116	2.60	513	106
2.90	530	103	2.50	530	116	2.60	513	106
2.70	528	109	2.60	534	114			
2.50	-	-	2.50	530	116	2.63	524	109
2.80	522	103	2.50	523	113			

Springfield.

COAL.				WATER.				COMMERCIAL.		
B.T.U.	Candle Power.	Yield.	B.T.U. Feet.	Gals.	B.T.U.	Factor.	Candle Power.	Per Cent. H₂O.	B.T.U.	Candle Power.
539	13.3	5.50	2,908	3.28	576	109	18.7	-	566	16.5
538	14.9	5.86	-	3.45	563	100	18.8	-	-	-
548	14.2	5.68	-	3.11	548	104	19.1	-	550	16.9
554	14.0	5.55	-	2.50	553	125	18.7	-	530	16.0
556	13.6	5.47	-	2.69	523	107	18.3	-	532	16.1
579	15.1	5.65	3,200	2.89	537	106	18.1	51.0	-	16.2
.559	15.1	5.55	3,102	2.80	509	99	18.6	56.0	522	16.7

Springfield — Continued.

COAL.				WATER.				COMMERCIAL.		
B.T.U.	Candle Power.	Yield.	B.T.U. Feet.	Gals.	B.T.U.	Factor.	Candle Power.	Per Cent. H₂O.	B.T.U.	Candle Power.
547	14.5	5.51	3,013	2.85	505	96	17.6	55.0	520	14.8
565	14.3	5.39	3,045	3.24	545	100	18.5	53.0	513	15.6
538	13.9	5.64	3,034	2.76	505	98	18.2	52.9	518	15.5
549	14.1	5.47	3,003	3.09	507	88	17.9	53.1	513	15.4
533	14.2	5.46	2,910	3.53·	533	90	19.0	53.0	519	15.3
542	14.7	5.33	2,887	3.23	523	93	19.0	53.5	524	15.6
545	14.9	5.32	2,897	3.08	536	101	19.2	51.0	519	14.9
555	15.1	5.24	2,908	2.95	542	106	18.9	52.0	519	15.5
550	14.8	5.40	2,969	3.20	542	100	18.8	52.8	515	15.2
549	13.5	5.47	3,002	3.22	593	115	17.8	50.8	517	13.8
535	13.0	5.26	2,814	3.19	536	98	18.1	57.7	520	14.2
544	14.3	5.49	2,983	3.01	531	101	18.3	56.5	522	15.3
553	14.8	5.52	3,050	3.16	541	100	18.3	60.8	533	15.9
554	14.6	5.10	2,825	3.32	548	99	18.6	64.1	560	14.7
557	13.8	5.03	2,801	3.05	577	111	19.2	61.8	521	14.6
535	13.4	5.16	2,761	3.09	527	97	18.8	63.2	518	14.5
520	12.9	5.11	2,654	3.12	553	105	19.0	61.6	515	14.4
537	14.0	5.03	2,702	3.32	548	91	18.9	57.5	526	15.0
512	11.9	5.19	2,657	3.40	547	97	18.8	43.1	521	14.6
563	14.3	5.09	2,865	3.12	575	112	19.1	51.4	531	15.0
560	14.2	5.22	2,923	3.27	567	106	18.5	52.7	527	15.5
532	13.9	4.97	2,644	2.89	543	108	18.8	53.5	528	15.7
549	13.9	5.07	2,782	3.12	540	101	18.7	52.4	550	15.3
541	13.2	5.12	2,769	3.00	550	107	18.8	62.3	524	14.3
541	13.2	4.86	2,629	3.15	540	100	18.1	60.4	521	14.6
525	8.4	5.63	2,955	3.29	535	100	18.2	56.9	518	13.9
519	7.0	5.58	2,896	2.95	533	103	18.2	55.7	520	14.0
505	7.5	5.81	2,934	3.04	524	98	17.8	54.7	519	13.8
508	7.0	5.75	2,911	3.14	533	98	18.5	55.8	520	14.2
505	7.1	5.68	2,868	3.18	539	99	19.3	57.9	520	14.5
–	–	5.75	–	3.15	–	–	–	57.3	–	–
497	6.9	6.17	3,056	3.06	538	98	18.7	60.0	529	14.9
505	7.4	5.72	2,888	3.02	521	97	18.1	56.1	532	14.2
530	8.3	5.73	3,036	3.12	·534	99	18.4	61.2	531	14.4
501	7.0	5.93	3,024	3.31	529	93	18.4	59.0	528	14.7
518	8.9	5.85	3,030	3.19	533	97	18.9	62.0	529	15.2

Springfield — Concluded.

COAL.				WATER.				COMMERCIAL.		
B.T.U.	Candle Power.	Yield.	B.T.U. Feet.	Gals.	B.T.U.	Factor.	Candle Power.	Per Cent. H₂O.	B.T.U.	Candle Power.
518	7.4	5.55	2,874	3.10	541	102	19.5	68.0	528	15.5
517	7.6	5.13	2,652	3.14	534	98	19.4	69.0	525	16.2
–	7.0	5.21	–	3.20	–	–	19.0	62.0	527	15.9
505	6.7	5.24	2,646	3.18	537	98	19.3	59.0	524	15.3
507	7.4	4.12	2,088	3.21	554	103	19.7	62.0	526	15.4
538	13.3	4.79	2,577	3.22	545	100	19.9	51.6	535	16.1
529	7.4	5.22	2,861	3.22	547	101	19.3	48.5	535	15.9
532	12.9	5.11	2,718	3.18	547	102	–	44.0	549	16.1
540	10.9	5.13	2,770	3.34	552	99	19.3	46.0	545	–
–	14.8	4.83	–	3.16	–	–	18.4	51.0	–	16.7
527	14.7	4.93	2,598	3.07	534	100	19.3	54.4	526	15.4
523	14.0	4.88	2.552	3.02	515	95	19.1	55.0	519	14.1
526	13.7	5.01	2,635	3.26	534	96	19.1	53.0	529	14.8
530	13.6	4.97	2,634	3.20	548	102	19.4	56.0	534	15.5
521	12.3	5.13	2,673	3.26	536	96	19.3	52.0	524	15.3
535	12.1	5.33	2,913	3.13	540	101	18.9	55.9	527	15.2

Suburban.

COMMERCIAL.		COMMERCIAL.		COMMERCIAL.		COMMERCIAL.	
B. T. U.	Candle Power.	B. T. U.	Candle Power.	B. T. U.	Candle Power.	B. T. U.	Candle Power.
555	16.5	534	14.7	531	13.6	533	14.5
548	14.7	535	14.5	521	11.9	534	14.8
535	13.5	537	14.7	524	12.7	538	14.3
519	12.8	540	14.1	530	13.4	533	15.0
521	13.5	540	14.4	541	14.5	537	15.1
531	14.5	548	14.6	524	14.1	538	13.2
538	13.8	525	13.4	545	14.8	531	14.4
529	13.6	534	14.4	551	15.4	538	14.8
533	14.0	541	15.5	539	14.8	538	14.8
532	14.8	548	14.7	546	14.9	537	15.9
537	13.8	530	14.9	547	15.2	532	14.2
528	14.9	524	14.3	539	15.5	535	14.4
520	15.0	528	15.3	543	15.3		
526	13.5	531	15.1	535	14.5		
532	14.9	535	–	534	14.3		

Taunton.

COAL.				WATER.				COMMERCIAL.		
B.T.U.	Candle Power.	Yield.	B.T.U. Feet.	Gals.	B.T.U.	Factor.	Candle Power.	Per Cent. H$_2$O.	B.T.U.	Candle Power.
623	14.5	4.85	3,021	–	–	–	–	–	–	–
620	15.4	4.82	2,990	–	–	–	–	–	–	–
613	14.4	5.05	3,093	–	–	–	–	–	–	–
598	13.1	5.10	3,050	–	–	–	–	–	–	–
619	14.3	4.79	2,964	–	–	–	–	–	–	–
616	14.8	4.72	2,908	–	–	–	–	–	–	–
605	13.9	5.08	3,073	–	–	–	–	–	–	–
609	13.6	5.19	3,161	–	–	–	–	–	–	–
616	13.5	4.85	2,988	–	–	–	–	–	–	–
603	13.5	4.77	2,878	–	–	–	–	–	–	–
602	13.3	4.85	2,923	–	–	–	–	–	–	–
615	13.8	4.76	2,926	–	–	–	–	–	–	–
613[1]	14.5[1]	4.86	2,978	3.09	–	–	–	4.7	613	14.5
627[1]	14.3[1]	4.92	3,084	4.10	–	–	–	0.9	627	14.3
–	–	5.01	–	–	–	–	–	7.4	598	13.9
573[1]	13.2[1]	5.17	2,962	2.92	–	–	–	2.0	573	13.2
595[1]	13.1[1]	4.64	2,762	–	–	–	–	0.0	595	13.1
603[1]	13.4	4.68	2,823	–	–	–	–	0.0	603	13.4
604[1]	12.9	4.80	2,899	–	–	–	–	0.0	604	12.9
612	13.4	4.85	2,967	–	–	–	–	0.0	612	13.4
629	14.7	4.85	3,052	–	–	–	–	–	–	–
586	12.4	4.71	2,759	–	–	–	–	–	–	–
605	12.8	4.67	2,825	–	–	–	–	–	–	–
624	14.7	4.78	2,984	–	–	–	–	–	–	–
617	–	4.82	2,974	–	–	–	–	–	–	–
618	13.6	4.67	2,885	–	–	–	–	–	–	–
616	14.0	4.86	2,994	–	–	–	–	–	–	–
625	13.1	4.73	2,958	–	–	–	–	–	–	–
605	13.8	5.06	3,059	–	–	–	–	–	–	–
623	14.7	4.88	3,042	–	–	–	–	–	–	–
635	14.6	4.77	3,027	–	–	–	–	–	–	–
621	14.5	5.00	3,104	–	–	–	–	–	–	–
609	13.5	5.18	3,153	–	–	–	–	–	–	–
608	14.5	4.99	3,039	–	–	–	–	–	–	–
601	14.3	5.14	3,087	–	–	–	–	–	–	–

[1] Returns of Btu and candle power are for commercial gas.

Taunton — Concluded.

COAL.				WATER.				COMMERCIAL.		
B.T.U.	Candle Power.	Yield.	B.T.U. Feet.	Gals.	B.T.U.	Factor.	Candle Power.	Per Cent. H_2O.	B.T.U.	Candle Power.
612	15.3	5.10	3,122	-	-	-	-	-	-	-
621	14.9	4.94	3,070	-	-	-	-	-	-	-
618	15.0	4.65	2,876	-	-	-	-	-	-	-
620	15.0	4.75	2,943	-	-	-	-	-	-	-
602	13.8	4.87	2,932	-	-	-	-	-	-	-
599	13.9	4.92	2,948	-	-	-	-	-	-	-
579	13.8	4.92	2,847	-	-	-	-	-	-	-
611	13.0	5.07	3,097	-	-	-	-	-	-	-
594	12.0	5.05	2,999	-	-	-	-	-	-	-
610	13.6	4.89	2,980	-	-	-	-	-	-	-

Worcester.

B.T.U.	Candle Power.	Yield.	B.T.U. Feet.	Gals.	B.T.U.	Factor.	Candle Power.	Per Cent. H_2O.	B.T.U.	Candle Power.
547	12.9	5.76	3,148	3.27	606	117	19.2	41.6	545	17.0
563	13.7	5.33	3,001	3.73	561	94	18.5	47.0	555	16.2
565	15.8	5.39	3,046	3.34	565	103	18.6	50.0	565	17.0
521	14.2	5.53	2,877	3.70	573	98	19.5	49.8	559	16.7
559	14.1	5.42	3,027	3.41	559	100	19.0	51.8	568	16.5
526	12.1	5.46	2,874	3.47	575	103	19.8	50.0	542	15.9
540	11.9	5.20	2,806	3.67	650	119	20.7	53.0	549	15.9
577	12.3	5.17	2,992	4.11	609	99	19.8	54.2	566	17.9
543	-	5.82	3,163	3.83	611	105	19.8	50.6	562	16.8
602	15.2	4.97	2,991	3.97	650	112	-	38.2	589	17.9
532	12.1	5.57	2,962	4.03	615	103	-	42.7	526	17.1
528	12.5	-	-	4.11	-	-	-	39.6	562	15.9
-	13.5	5.14	-	4.19	-	-	-	44.9	584	16.5
563	12.9	4.78	2,688	5.09	640	91	23.4	52.2	580	15.2
555	10.6	5.44	2,999	4.47	645	101	23.0	46.1	579	15.2
534	10.9	4.90	2,615	4.15	639	106	-	40.1	580	15.3
-	-	5.45	-	4.13	-	-	-	46.0	551	15.6
586	11.1	5.27	-	4.46	671	107	-	45.5	575	15.5
523	10.9	5.32	2,781	4.11	622	103	-	50.7	561	14.7
499	10.1	5.89	2,695	3.79	609	106	24.1	49.0	542	14.6
474	9.4	5.73	2,713	4.35	616	97	24.1	49.1	533	14.4
524	11.4	4.93	2,591	4.60	640	98	23.3	48.0	573	15.6
444	10.5	5.16	2,293	4.12	620	102	23.3	49.1	538	15.4
413	10.9	5.14	2,122	4.00	608	110	24.0	42.6	537	13.4

Worcester — Concluded.

COAL.				WATER.				COMMERCIAL.		
B.T.U.	Candle Power.	Yield.	B.T.U. Feet.	Gals.	B.T.U.	Factor.	Candle Power.	Per Cent. H₂O.	B.T.U.	Candle Power.
523	11.2	5.09	2,662	4.21	632	103	23.8	37.1	561	15.2
524	9.1	4.89	2,721	4.36	632	100	22.5	43.7	555	14.9
594	12.4	4.74	2,817	3.85	628	109	21.6	42.7	580	16.0
510	11.5	4.39	2,241	3.70	626	112	21.1	48.0	556	15.9
519	9.3	5.02	2,608	4.19	644	106	19.2	38.2	552	14.7
488	12.5	5.05	2,463	3.84	618	107	23.8	29.6	552	15.0
537	11.5	4.92	2,642	4.04	611	101	22.3	38.4	561	15.3
479	12.3	4.85	2,323	3.74	613	108	23.6	34.2	561	14.9
508	10.1	4.97	2,524	3.90	620	106	23.2	31.0	561	14.4
545	12.2	4.65	2,536	3.85	627	109	24.6	39.7	563	14.9
541	12.1	5.11	2,764	3.96	632	108	27.1	33.2	561	14.4
531	11.9	5.11	2,714	4.42	615	95	23.7	29.8	549	13.9
521	12.4	5.17	2,693	4.09	617	102	22.8	37.5	567	13.9
584	14.4	5.69	3,320	3.15	595	118	22.2	60.2	560	16.1
507	13.4	5.69	2,886	3.04	522	97	20.0	59.2	539	14.9
580	12.9	5.66	3,281	3.00	570	114	19.2	64.5	555	15.1
546	12.7	5.70	3,214	2.75	542	112	17.3	56.9	551	15.5
559	12.2	5.59	3,125	2.54	524	112	15.9	65.3	536	15.3
561	11.8	5.60	3,139	2.90	535	105	15.6	57.3	537	14.1
561	12.2	5.79	3,245	2.70	515	104	15.4	61.6	530	15.3
566	12.1	5.70	3,226	3.05	556	105	16.1	55.3	539	15.6
545	12.8	5.54	3,016	2.81	538	107	17.7	57.7	543	15.1
549	12.2	5.52	3,029	2.84	533	106	19.1	54.7	540	12.5
561	12.6	5.51	3,092	2.34	498	110	13.4	33.6	559	14.7
567	12.5	5.53	3,135	2.37	481	100	8.6	39.9	535	14.4
573	14.1	5.45	3,107	1.88	479	122	9.9	38.0	536	14.4
559	11.8	5.46	3,054	2.95	495	90	13.8	36.0	536	14.4
556	12.2	5.42	3,015	2.47	495	103	10.4	37.2	540	13.3
521	12.2	5.45	2,842	2.69	505	100	13.7	42.9	535	12.9
551	12.9	5.52	3,042	2.56	489	98	12.3	41.1	538	12.4
559	10.4	5.34	2,986	2.62	494	109	11.2	41.9	553	14.4
561	10.6	5.51	3,093	2.52	493	101	10.8	47.9	539	12.6
541	10.1	5.59	3,026	3.15	532·	98	16.0	50.8	540	14.1
571	10.5	5.57	3,178	3.19	546	101	–	49.2	557	14.9
555	9.2	5.57	3,089	2.87	528	104	15.4	48.5	543	15.0
571	9.8	5.54	3,165	2.94	545	107	15.4	52.1	550	14.9
540	11.9	5.32	2,882	3.52	579	104	16.0	46.1	553	15.0

STATE INSPECTIONS.

COMPANY.	STATE TESTS.		COMPANIES' TESTS FOR WEEK, INCLUDING DATE OF STATE TEST.			
	Date.	B. T. U.	Test Station, Average.	Maximum.	Minimum.	Works, Average.
Attleboro, ...	Oct. 25	603	605	-	-	-
	Nov. 15	619	607	627	590	-
	Dec. 5	612	611	-	-	-
	Dec. 27	609	611	-	-	-
	Jan. 16	613	607	617	598	-
	Feb. 20	605	604	616	597	-
	Mar. 21	629	626	632	617	-
	Apr. 24	616	621	633	609	-
	Aug. 2	630	622	631	597	-
Boston, ...	Aug. 1	602	595	-	-	590
	Aug. 7	586	581	-	-	583
	Aug. 16	584	585	-	-	592
	Aug. 23	592	582	593	571	576
	Aug. 30	561	564	-	-	565
	Sept. 20	563	564	-	-	558
	Sept. 26	569	576	585	569	559
	Oct. 13	565	576	592	564	558
	Oct. 27	577	583	-	-	573
	Nov. 4	587	584	-	-	578
	Nov. 21	595	595	-	-	590
	Dec. 5	596	594	-	-	580
	Dec. 21	585	584	-	-	582
	Jan. 9	574	572	581	558	566
	Jan. 18	581	581	590	574	570
	Feb. 1	573	573	581	559	566
	Mar. 3	570	572	576	566	575
	Mar. 18	578	578	584	571	576
	Mar. 25	566	568	572	564	575
	Apr. 18	591	582	585	579	575
	Apr. 25	583	584	593	575	581
	May 1	585	589	599	583	592
	May 15	619	586	597	576	580
	June 12	583	589	596	582	587
	June 27	597	593	597	586	597
	July 6	600	599	602	595	595

STATE INSPECTIONS — *Continued.*

COMPANY.	STATE TESTS.		COMPANIES' TESTS FOR WEEK, INCLUDING DATE OF STATE TEST.			
	Date.	B. T. U.	Test Station, Average.	Maximum.	Minimum.	Works, Average.
Boston — *Con.* . . .	July 25	590	589	598	583	584
	Aug. 17	569	571	576	568	569
	Sept. 14	559	559	568	554	558
	Sept. 21	570	564	571	556	564
Brockton, ...	Aug. 31	558	567	591	556	–
	Sept. 21	563	565	–	–	570
	Oct. 11	547	553	–	–	567
	Nov. 1	549	555	561	547	569
	Nov. 10	560	553	–	–	575
	Nov. 28	553	559	–	–	577
	Dec. 15	553	559	–	–	573
	Jan. 23	543	542	545	538	558
	Feb. 27	555	548	555	543	565
	Mar. 28	567	565	571	554	582
	Apr. 25	553	551	556	546	571
	July 5	550	550	554	546	569
	July 31	550	550	555	542	577
Cambridge, ...	Aug. 5	567	570	–	–	576
	Aug. 11	560	566	588	560	572
	Aug. 16	561	555	–	–	560
	Sept. 16	552	558	–	–	560
	Sept. 22	560	559	–	–	568
	Oct. 11	554	556	–	–	561
	Nov. 1	569	569	–	–	566
	Nov. 11	582	577	–	–	577
	Nov. 17	583	581	–	–	577
	Nov. 23	576	587	591	581	591
	Dec. 6	596	592	–	–	590
	Dec. 30	579	575	–	–	577
	Jan. 5	572	579	584	574	577
	Jan. 20	575	576	587	568	570
	Feb. 6	567	574	577	570	578
	Mar. 13	573	572	578	564	568
	Apr. 20	577	569	577	563	564
	July 28	553	562	563	560	564
	Sept. 28	558	556	558	552	557

STATE INSPECTIONS — *Continued.*

COMPANY.	STATE TESTS.		COMPANIES' TESTS FOR WEEK, INCLUDING DATE OF STATE TEST.			
	Date.	B. T. U.	Test Station, Average.	Maximum.	Minimum.	Works, Average.
Charlestown, . . .	Feb. 27	571	612	–	–	–
	Mar. 6	630	617	–	–	–
	Apr. 7	613	619	–	–	–
	Apr. 17	611	621	–	–	–
	May 1	602	615	–	–	–
	June 13	600	606	–	–	–
	Sept. 29	609	609	615	602	–
East Boston, . . .	Oct. 30	559	548	562	532	–
	Nov. 8	539	548	556	528	–
	Nov. 14	535	542	–	–	–
	Dec. 1	537	534	–	–	–
	Dec. 21	546	535	550	524	–
	Jan. 5	527	542	571	526	–
	Jan. 24	530	527	532	519	–
	Feb. 17	526	526	539	513	–
	Mar. 25	543	536	545	524	–
	Apr. 18	519	527	537	521	–
	Apr. 26	510	515	534	508	–
	June 12	520	523	580	486	–
	June 25	499	–	–	–	–
	July 7	532	534	544	522	–
	July 25	526	543	558	526	–
	Sept. 14	522	529	536	522	–
Fall River, . . .	July 31	570	548	558	538	561
	Aug. 11	556	554	–	–	552
	Sept. 21	548	553	–	–	555
	Oct. 18	549	560	569	545	557
	Oct. 31	574	563	566	557	554
	Nov. 10	557	556	–	–	556
	Nov. 16	570	562	570	558	554
	Nov. 29	567	556	565	549	553
	Dec. 15	545	547	–	–	548
	Dec. 29	557	554	–	–	548
	Jan. 12	566	559	565	550	555
	Jan. 23	568	560	566	554	556

STATE INSPECTIONS — *Continued.*

COMPANY.	STATE TESTS.		COMPANIES' TESTS FOR WEEK, INCLUDING DATE OF STATE TEST.			
	Date.	B. T. U.	Test Station, Average.	Maxi- mum.	Mini- mum.	Works, Average.
Fall River — *Con.*	Feb. 7	562	561	569	551	564
	Feb. 23	566	563	566	557	565
	Mar. 20	555	548	553	541	438
	Mar. 28	561	552	561	546	539
	May 2	540	535	542	522	-
	July 11	551	557	562	546	542
	Sept. 26	540	542	556	532	536
Fitchburg,	Aug. 7	571	551	582	522	539
	Aug. 23	559	532	542	523	547
	Sept. 16	540	550	565	530	560
	Oct. 10	522	536	546	523	534
	Oct. 16	519	532	541	523	533
	Nov. 7	538	531	-	-	532
	Nov. 22	565	558	572	533	571
	Dec. 13	566	564	-	-	568
	Dec. 21	563	561	-	-	577
	Jan. 26	568	553	567	547	561
	Feb. 13	560	561	573	551	555
	Apr. 17	550	549	556	538	542
	June 9	537	540	547	523	540
	Aug. 8	552	548	559	536	545
Haverhill,	Aug. 18	564	559	573	550	-
	Sept. 15	549	547	-	-	-
	Sept. 27	567	551	557	541	-
	Oct. 23	562	559	-	-	-
	Nov. 9	570	565	-	-	557
	Nov. 23	566	563	-	-	554
	Dec. 28	557	555	-	-	536
	Jan. 25	549	546	554	534	546
	Mar. 2	558	551	559	545	544
	Mar. 14	548	547	554	538	-
	Apr. 26	555	552	554	546	-
	July 27	540	535	544	530	-
	Sept. 13	543	-[1]	-	-	-

[1] No tests made.

State Inspections — *Continued.*

COMPANY.	STATE TESTS.		COMPANIES' TESTS FOR WEEK, INCLUDING DATE OF STATE TEST.			
	Date.	B. T. U.	Test Station, Average.	Maximum.	Minimum.	Works, Average.
Holyoke, ...	Aug. 11	545	571	598	552	–
	Aug. 17	578	579	–	–	–
	Aug. 24	600	–	–	–	–
	Sept. 19	608	–	–	–	–
	Nov. 16	577	–	–	–	–
	Dec. 6	565	581	605	568	–
	Dec. 30	577	–	–	–	–
	Jan. 12	568	–	623 [1]	586 [1]	601
	May 17	560	544	560	501	560
	June 8	556	577	600	549	591
	Aug. 30	580	577	576	566	–
Lowell, ...	Sept. 15	589	588	–	–	589
	Oct. 23	594	595	–	–	597
	Nov. 7	595	600	602	598	602
	Nov. 22	570	604	607	602	603
	Dec. 12	579	583	–	–	575
	Dec. 29	556	579	583	581	577
	Jan. 19	588	588	591	586	585
	Jan. 30	582	577	585	569	580
	Mar. 2	568	573	578	568	570
	Apr. 24	580	585	588	583	571
	June 15	575	586	590	582	571
	July 24	593	594	610	583	586
	Aug. 30	583	582	582	578	–
	Sept. 11	574	573	573	573	573
Lynn, ...	Sept. 8	577	578	–	–	575
	Sept. 19	570	565	575	556	550
	Sept. 27	569	566	–	–	567
	Oct. 3	567	–	–	–	560
	Oct. 17	550	549	–	–	547
	Oct. 31	544	549	572	537	544
	Nov. 14	562	566	577	561	573
	Dec. 1	577	564	572	560	574
	Dec. 28	571	560	566	544	574

[1] Taken at works station.

STATE INSPECTIONS — *Continued.*

COMPANY.	STATE TESTS.		COMPANIES' TESTS FOR WEEK, INCLUDING DATE OF STATE TEST.			
	Date.	B. T. U.	Test Station, Average.	Maximum.	Minimum.	Works, Average.
Lynn — *Con.*	Jan. 10	560	553	557	548	570
	Jan. 31	571	560	562	558	565
	Mar. 1	560	550	560	537	559
	Mar. 14	570	568	573	560	573
	Apr. 17	550	557	560	552	558
	May 3	555	552	556	548	554
	June 13	539	536	538	535	541
	July 26	545	540	544	536	535
	Sept. 21	547	539	542	533	546
Malden,	Aug. 18	557	–	–	–	560
	Aug. 24	542	–	–	–	564
	Sept. 9	535	–	–	–	544
	Sept. 20	526	533	542	519	534
	Oct. 3	528	533	–	–	532
	Oct. 5	537	545	561	537	532
	Oct. 30	544	546	–	–	543
	Nov. 14	552	547	553	544	540
	Nov. 23	534	538	–	–	536
	Dec. 21	534	542	561	533	540
	Jan. 9	557	548	551	539	544
	Mar. 2	539	531	541	519	536
	Mar. 24	546	543	550	532	550
	Apr. 20	538	531	539	516	530
	May 5	541	544	553	534	544
	July 27	537	533	537	530	535
	Aug. 31	529	532	546	526	531
New Bedford,	Aug. 16	590	578	591	565	597
	Aug. 25	567	560	568	552	578
	Sept. 22	568	558	570	547	587
	Oct. 25	567	569	–	–	–
	Oct. 31	551	554	–	–	–
	Nov. 16	565	556	563	546	–
	Nov. 29	576	570	578	557	576
	Dec. 14	586	590	594	571	596
	Dec. 29	573	578	–	–	589

STATE INSPECTIONS — *Continued.*

COMPANY.	STATE TESTS.		COMPANIES' TESTS FOR WEEK, INCLUDING DATE OF STATE TEST.			
	Date.	B. T. U.	Test Station, Average.	Maximum.	Minimum.	Works, Average.
New Bedford — *Con.*. .	Jan. 12	533	552	577	534	561
	Jan. 25	568	560	565	551	570
	Feb. 25	569	562	574	552	566
	Mar. 20	571	565	567	563	570
	May 2	543	544	558	535	558
	July 11	543	565	578	558	570
	Sept. 26	545	552	562	542	551
Old Colony, . . .	July 29	567	557	563	536	-
	Aug. 9	547	544	-	-	-
	Sept. 9	536	528	538	518	-
	Sept. 13	551	530	544	516	-
	Oct. 13	509	509	-	-	-
	Oct. 24	515	518	-	-	-
	Nov. 16	523	534	-	517	-
	Nov. 28	519	530	543	518	-
	Dec. 29	521	529	529	514	-
	Jan. 10	528	-	538	515	527
	Mar. 13	524	-	-	-	526
	Sept. 6	513	514	525	507	-
Springfield, . . .	Aug. 16	548	550	-	-	555
	Aug. 25	538	530	-	-	559
	Sept. 20	517	520	-	-	535
	Oct. 18	536	519	521	516	533
	Nov. 17	524	515	516	514	531
	Dec. 7	521	522	-	-	524
	Dec. 29	523	-	-	-	-
	Jan. 23	535	526	531	514	531
	Feb. 23	530	528	539	521	543
	Mar. 8	522	524	532	515	545
	Mar. 16	519	521	528	515	534
	Mar. 23	523	518	527	510	524
	Apr. 18	532	520	523	518	520
	Apr. 25	534	-	-	-	-
	May 17	522	528	537	520	522
	June 8	534	529	534	519	524

STATE INSPECTIONS — *Continued.*

COMPANY.	STATE TESTS.		COMPANIES' TESTS FOR WEEK, INCLUDING DATE OF STATE TEST.			
	Date.	B. T. U.	Test Station, Average.	Maximum.	Minimum.	Works, Average.
Springfield — *Con.*	July 19	546	537	540	531	535
	Aug. 10	524	530	562	512	545
	Sept. 5	501	515	526	504	519
Suburban,	Aug. 30	553	555	-	-	-
	Sept. 19	526	519	529	514	-
	Oct. 17	532	538	-	-	-
	Nov. 21	529	528	-	-	-
	Jan. 18	545	540	544	534	-
	Mar. 23	541	531	541	524	-
	Sept. 28	529	532	537	528	-
Taunton,	Sept. 22	619	-	-	-	-
	Oct. 18	617	620	-	-	-
	Oct. 25	616	613	-	-	-
	Nov. 28	609	609	-	-	-
	Dec. 27	625	615	621	605	-
	Feb. 27	607	595	605	583	-
	Mar. 4	607	595	605	583	-
	Apr. 25	654	624	649	615	-
	July 5	642	608	622	598	-
	Aug. 3	627	618	633	603	-
	Sept. 18	625	611	623	597	-
Worcester,	July 21	563	-	-	-	-
	Aug. 2	552	545	555	531	545
	Aug. 11	551	555	-	-	575
	Aug. 16	567	565	-	-	584
	Aug. 25	559	559	-	-	569
	Sept. 20	564	-	-	-	-
	Oct. 10	606	589	623	550	617
	Oct. 27	577	562	578	509	575
	Nov. 3	541	584	598	568	-
	Nov. 17	588	575	588	557	589
	Dec. 14	548	561	591	543	571
	Dec. 29	529	533	-	-	548
	Jan. 11	539	538	548	531	548
	Jan. 23	567	561	573	555	571

STATE INSPECTIONS — *Concluded.*

COMPANY.	STATE TESTS.		COMPANIES' TESTS FOR WEEK, INCLUDING DATE OF STATE TEST.			
	Date.	B. T. U.	Test Station, Average.	Maximum.	Minimum.	Works, Average.
Worcester — *Con.* . .	Feb. 25	520	552	579	523	558
	Mar. 16	552	554	567	546	561
	Apr. 23	574	560	582	542	575
	May 4	544	539	548	529	540
	May 17	555	551	559	543	554
	June 21	545	539	555	525	557
	July 19	541	540	547	–	536
	Sept. 5	528	537	547	526	540
	Sept. 10	565	550	560	543	557
	Sept. 18	550	547	558	542	543

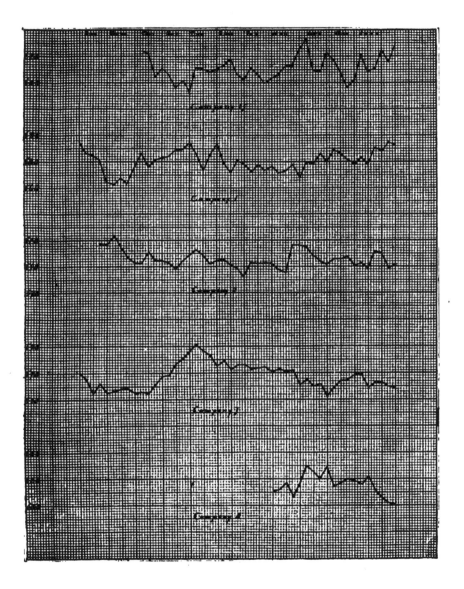

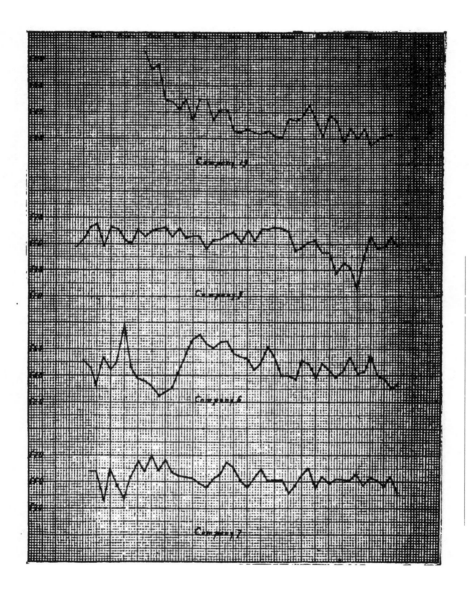

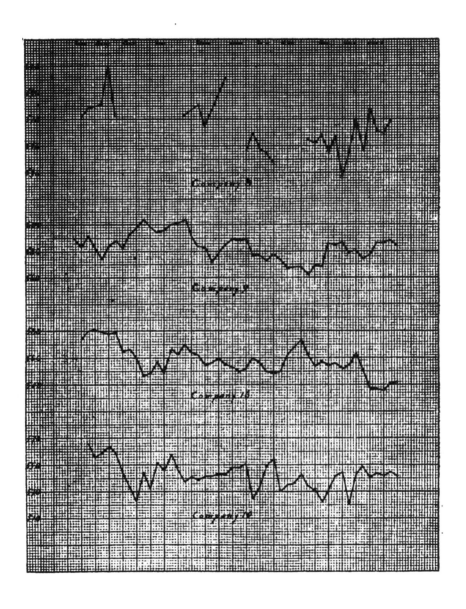

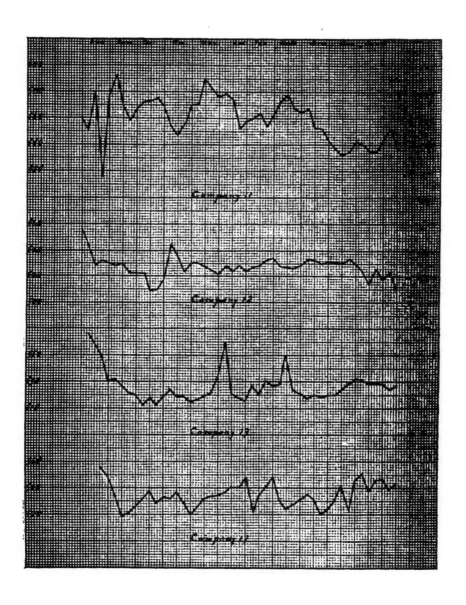

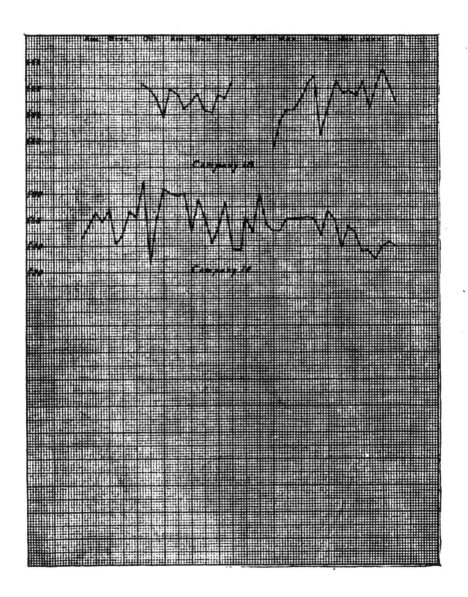

APPENDIX H.

BURNER TESTS.

The following are tables, with their averages, showing variations in open burner candle powers compared with results obtained in official tests with Suggs' Argand burners of suitable size (D, E and F) and appropriate chimney ($1\frac{7}{8}''$ x 6'' and $1\frac{7}{8}''$ x 7''):—

TABLE A. — *Water Gas, 1910.*

Argand.	Open.	Argand.	Open.	Argand.	Open.
20.82	20.28	21.40	21.30	20.27	20.66
20.40	20.82	20.66	21.55	19.30	17.21
18.30	13.66	20.55	20.47	17.60	15.40
20.89	23.23	20.44	16.61	14.61	10.53
20.48	19.15	19.72	19.64	20.94	20.25
18.33	17.84	20.10	17.98	20.65	19.34
20.80	19.40	19.95	19.78	20.74	19.02
19.59	18.58	20.70	18.80	19.84	19.51

TABLE B. — *Water Gas, 1912.*

Argand.	Open.	Argand.	Open.	Argand.	Open.
19.58	22.20	18.97	20.41	20.46	20.82
20.22	22.02	18.94	22.52	19.03	23.68
20.63	20.71	21.92	22.01	19.83	20.15
19.47	19.81	20.10	23.40	20.06	21.76
19.89	21.50	21.75	23.66		

TABLE C. — *Water Gas, 1912.*

Argand.	Open.	Argand.	Open.	Argand.	Open.
20.51	18.55	20.81	19.18	18.05	16.34
23.16	16.45	18.00	17.40	20.17	19.55
22.16	21.64	18.50	17.31	24.36	22.14
19.67	17.90	20.42	18.77	19.26	19.03
21.56	19.52	20.31	19.12	20.42	18.64
21.74	20.33	20.17	16.43		
19.73	19.03	18.99	16.75		

TABLE D. — *Coal Gas, 1910.*

Argand.	Open.	Argand.	Open.	Argand.	Open.
17.09	16.24	14.50	11.54	16.94	13.28
17.42	16.75	16.09	14.50	16.10	13.50
16.54	13.16	16.88	16.01	15.90	14.00
16.49	12.19	16.21	15.64	16.63	15.31
17.21	15.14	17.08	16.25	17.35	14.40
16.90	15.20	17.13	15.13	16.68	16.21
16.27	13.52	17.36	15.18	17.11	15.76
15.86	14.99	16.76	15.47	16.39	14.18
16.69	13.09	17.03	16.43	16.71	14.56
16.78	15.20	17.18	15.29	16.43	14.89

TABLE E. — *Coal Gas, 1912.*

Argand.	Open.	Argand.	Open.	Argand.	Open.
16.91	14.80	16.66	14.77	17.47	13.27
17.12	16.20	16.60	15.82	16.79	13.32
16.77	15.13	16.14	13.90	16.76	14.53
15.86	14.74	17.26	13.33		

TABLE F. — *Mixed Gas, 1910.*

Argand.	Open.	Argand.	Open.	Argand.	Open.
17.90	15.61	18.62	16.98	18.70	17.41
17.43	16.06	17.10	17.60	19.01	19.35
18.10	16.60	18.51	17.03	19.10	18.56
17.09	16.35	16.58	17.00	18.43	15.09
18.33	15.87	17.11	17.83	19.17	18.59
18.00	15.78	17.16	16.80	18.30	15.90
17.18	14.87	18.32	16.80	17.77	14.60
18.32	19.22	18.27	16.79	16.39	13.69
17.10	13.10	16.61	16.92	16.11	11.65
17.65	16.84	17.64	15.50	16.19	13.97
18.10	16.60	17.60	14.50	18.80	18.60
17.69	15.08	18.20	16.31	17.20	14.26
17.62	17.29	18.02	16.86	17.98	16.03
17.78	16.64	17.70	16.40	18.20	15.56
18.10	17.31	16.66	15.92	17.87	18.00
18.62	15.58	17.88	17.32	17.90	17.00
17.76	14.40	16.83	12.82	17.85	16.35
17.44	16.23	19.06	18.26		
18.19	15.22	19.13	18.87		

TABLE G. — *Mixed Gas, 1912.*

Argand.	Open.	Argand.	Open.	Argand.	Open.
17.43	16.43	18.63	16.62	16.22	12.51
18.17	17.63	18.53	16.49	16.24	13.24
16.62	15.82	16.19	12.10	17.83	16.10
16.37	15.55	19.65	18.68	17.45	14.31
17.24	17.07	17.14	15.41	18.05	18.02
18.40	17.13	19.70	17.09	18.43	18.24
18.03	16.03	18.24	14.07	17.75	15.68
18.31	16.25	17.43	13.09		

TABLE H. — *Plants not under Heat Unit Basis for Year ending 1916.*

COAL GAS.		WATER GAS.		MIXED GAS.	
Argand.	Open.	Argand.	Open.	Argand.	Open.
16.4	13.3	16.9	8.5	17.8	14.1
16.5	11.2	18.2	10.8	16.3	10.7
14.8	11.1	16.1	8.8	16.6	9.8
16.4	14.5	14.5	8.4	16.5	13.9
16.5	13.4	17.5	11.1	16.0	13.2
16.8	10.5	16.9	12.0	18.2	12.4
16.6	10.5	16.9	10.4	18.3	10.9
15.6	12.9	16.5	9.3	16.2	11.8
15.2	11.1	15.2	8.5	16.2	9.2
16.2	14.2	18.8	9.9	16.1	11.8
16.10	12.27	16.1	8.0	16.6	14.4
		16.1	8.1	17.2	14.9
		16.8	9.48	16.83	12.25

TABLE I. — *Plants under Heat Unit Basis, 1916.*

WATER GAS.			COAL GAS.		
B. T. U.	Argand.	Open.	B. T. U.	Argand.	Open.
522.6	14.5	7.9	609.3	13.8	9.2
518.3	12.5	5.3	612.0	16.3	13.1
566.3	16.4	12.1	603.4	16.4	11.8
535.9	14.47	8.43	608.2	15.50	11.37

TABLE I — *Concluded.*

B. T. U.	Argand.	Open.	B. T. U.	Argand.	Open.
596.2	17.2	10.6	566.0	14.5	8.9
553.2	13.7	8.1	577.1	16.0	13.8
553.0	12.4	7.9	565.0	15.4	12.8
581.9	15.2	10.3	579.0	14.7	11.5
582.5	15.1	13.4	569.8	15.9	13.0
576.2	15.6	11.1	551.9	15.0	8.6
596.1	16.7	12.4	534.1	14.4	8.8
535.1	14.3	7.1	585.8	18.9	9.7
570.4	14.5	8.2	524.1	13.5	7.0
567.0	14.2	10.0	520.8	15.7	10.6
545.0	14.1	8.8	588.0	16.5	11.3
556.9	14.6	9.5	548.0	14.5	10.0
565.3	14.3	9.7	563.9	15.08	10.12

Lightning Source UK Ltd.
Milton Keynes UK
UKHW012020021218
333216UK00014B/2378/P